Advanced Engine Development at Pratt & Whitney

The Inside Story of Eight Special Projects, 1946–1971

Dick Mulready

Society of Automotive Engineers, Inc.
Warrendale, Pa.

Library of Congress Cataloging-in-Publication Data

Mulready, Dick (Richard C.), 1925–
    Advanced engine development at Pratt & Whitney : the inside story of
eight special projects, 1946–1971 / Dick Mulready.
        p. cm.
    Includes index.
    ISBN 0-7680-0664-3
        1. Jet engines—United States—Design and construction—History.
        2. Rocket engines—Design and construction—History.
        3. Pratt & Whitney Aircraft Company—History. I. Title.

TL709 .M85 2001
621.43'5'0973—dc21   00-051614

Copyright © 2001   Society of Automotive Engineers, Inc.
                    400 Commonwealth Drive
                    Warrendale, PA 15096-0001 U.S.A.
                    Phone: (724) 776-4841
                    Fax: (724) 776-5760
                    E-mail: publications@sae.org
                    http://www.sae.org

ISBN 0-7680-0664-3

SAE Order No. R-252

John Chamberlain's genius and generous spirit
made possible many of the breakthroughs described in this work.

Contents

Preface

This book celebrates the wonderful projects on which we worked at Pratt & Whitney during the almost magical quarter century bounded by World War II and the competition to develop the Space Shuttle engine. It was a time when going to work was fun, with a new challenge arising almost daily. The feeling of loyalty, both upward and downward within our company, was palpable. The work was pioneering, never having been done in the past, and it was very exciting. Some of the work has never been described until this book because of stringent security classifications that are now lifted.

Through a string of good luck, I became the manager of a series of technically advanced projects that are covered sequentially in this book. The story comes mostly from my memory but has been corroborated by several people who were present when the events occurred. This book is about the almost unbelievable engines and the dedicated group of people who made the engines real. Because gaps exist in my memory, I cannot remember the names of all of the people involved; however, I can see their faces clearly in my mind.

The roster for our team was not constant during the 25 years covered in this book. Instead, it ebbed and flowed according to the particular skills needed for the projects. It is interesting how many people returned to the advanced work as time passed. Most of these unique projects were not the daily "bread and butter" for Pratt & Whitney and thus were free from much of the survival pressure that typically surrounds that work. Instead, they were driven by the challenge of attempting things that had never been done. The trust among the people involved was clearly visible, and the complete lack of ego made the team almost unstoppable.

An example was the birth of the RL10 rocket engine, which was the first use of liquid hydrogen fuel in an operational engine. The thermodynamic cycle was bootstrap in that energy rejected to the cooling jacket was used to drive the turbopumps, and it had never been tried on a rocket engine. Nothing was wasted. The test facility, which is unique, simulated starting and operation in space with two-stage steam ejectors and a convergent/divergent diffuser. The engine had a 40-to-1 area ratio exhaust nozzle, higher than ever employed in the past. With all of these new elements, the engine was running at full duration within nine months of the time that the first scoop of sand had been dug for the test site. Such rapid and successful progress is possible only with an experienced and dedicated group whose goals are solely to make something work.

As with most endeavors that search new ground, not all of the projects yielded useful products. Two lasting discoveries that came from the work of the group were the RL10 hydrogen rocket engine, which has been used to launch most large satellites over the past half-century, and the development of the technology for the high-pressure staged combustion rocket engine used in the Space Shuttle.

The last challenge for the group was the genesis of the staged combustion, high-pressure rocket, which became the basis for the Space Shuttle engine. This work, which spanned most of a decade, is covered in Chapters 5 and 6, and it represents almost one-half billion of today's dollars. Using the guise of a competition, NASA conspired to take this technology from us and give it to a competitor. Such action was not illegal because we had been forced to surrender all rights to the data as a condition for entering the competition. Although it does not happen often in the real world, the ruse backfired almost immediately in this case. For the most critical items, NASA was forced to return to us, and this story unfolds in the final chapter of this book.

Acknowledgments

I would like to thank the following people for their efforts in reviewing or contributing to this work:

Bob Abernethy	Tom Kmiec
Dave Bogue	Pete Manz
Norm Bott	Frank McAbee
John Chamberlain	Pete Mitchell
Bill Creslein	Don Riccardi
Dick Coar	Ed Pinsley
Jerry Cuffe	Dan Sims
Lou Emerson	Frank Williams
Hal Gibson	George Zewski
Carl Kah	

CHAPTER 1

Ramjets—The Early Days at the Research Laboratory

The United Aircraft Research Laboratory (UARL) was a magical world in the fall of 1946. At that time, United Aircraft Corporation consisted of four major divisions: Pratt & Whitney Aircraft (engines), Chance Vought Aircraft (fighter planes), Hamilton Standard Propeller (propellers and controls), and Sikorsky Aircraft (helicopters). The United Aircraft Research Laboratory supported all of the manufacturing divisions.

There, a group of bright young people, most of whom were newly hired, pursued advanced technologies in thermodynamics and aerodynamics, which are the fundamentals of the business of engines and aircraft. All of the work was very advanced and eventually included other fields such as lasers and fusion. The facilities were the best and the newest, and the computer laboratory was being formed for technical as well as business calculations. It became one of the first users of the new large IBM machines.

The place was nicknamed "Wind Tunnel" and was located on the airport. The large concrete structure dominates the building and is the first thing you see. The philosophy of the corporation was to provide all facilities needed. At the time, it was the largest privately owned wind tunnel in the world. Figure 1.1 is a diagram of the Wind Tunnel.

This imposing test structure was built of 17,000 tons of concrete and steel. It is more than 600 feet long and 200 feet wide, with walls in some places as thick as two and one-half feet. An innovative design permitted the test sections and corresponding diffusers, which were mounted on railroad tracks, to be rolled into position to allow testing at a variety of speed ranges.

Subsonic testing had been underway for several months. Shakedown testing of the high-speed section of the tunnel began, and a major glitch developed because the attained velocity was significantly lower than had been expected.

Separation is a dirty word to those who deal with fluids that flow, and the tunnel diffuser was separating. This tunnel is the closed-loop variety, and the air circulates from the test section,

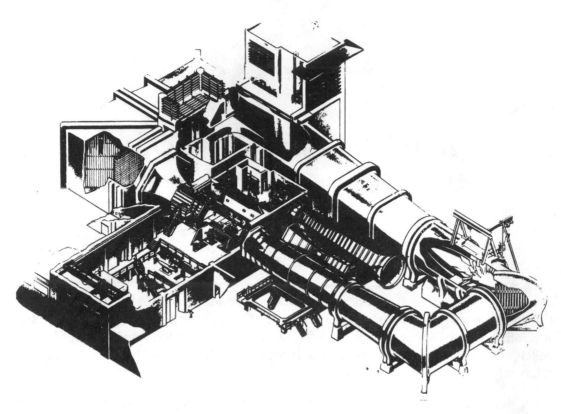

Figure 1.1   The Wind Tunnel.   (Courtesy of Pratt & Whitney)

through the diffuser, around two corners, through the fan, around two more corners, and back to the test section. The problem was that the diffuser pressure recovery was too low, and the tunnel needed something to keep the flow stuck to the inside walls of the diffuser.

"Henk" (Hendrick) Bruynes, who worked in the analytical section of the Research Laboratory, invented the "Vortex Generator," which proved to be the solution. It was a ring of eight short airfoils, cantilevered from the walls of the diffuser, with alternating clockwise and counterclockwise angles of attack. The airfoils generated tip vortices, which energized the boundary layer and prevented separation. So equipped, the tunnel achieved very close to sonic speed. Anyone who flies has seen this invention used on the wings of Boeing airplanes. A row of vortex generators at the proper location along the wing delays separation and improves lift. Fifty years later, the tunnel remains in operation and works effectively, primarily for the Sikorsky helicopter division.

In that day, one of the tests done in the large tunnel was that of a large-scale helicopter rotor model in a flying environment. A movie camera was located on the hub, facing outward along the blade. The view of the gyrations through which a blade goes, each revolution, is etched in my memory forever. The thought invariably comes to mind whenever I take off in a helicopter.

At that time, the director of research was Frank Caldwell, famous for inventing the controllable pitch aircraft propeller. Frank was scheduled to retire early the following year, to be succeeded by his assistant, John Lee. John was a tall and professorial type, with thick glasses and a bushy mustache. He was liked universally, and his presence added to the collegial aura. Some say that John had the habit of reading reports while he drove to work. With his thick glasses, John must have possessed reflexes of lightening, not evident from his appearance, to be able to drive that way.

Several years prior to this time, when he was a new aeronautical engineer from Massachusetts Institute of Technology (MIT), Stout, the legendary seat-of-the-pants leader of the Ford Airplane Company, had hired John Lee as his technical staff. At that time, Ford built a single-engine monoplane, and Stout was feeling competitive pressure from the new Fokker three-engine airplane. Stout called Lee into his office one day and announced that they would build a three-engine airplane. Lee suggested that he should go to the wind tunnel at MIT and start tests for the new wing. Stout said, "Never mind a new wing. Just cut notches for the additional engines in the old wing." Per Stout's directive, the airplane was built that way. When the test pilot took off for the first flight, it became apparent immediately that the wing was extremely overloaded. The pilot gingerly landed the plane and stormed off in search of Stout. The hangar burned down that night.

Before the plane could be rebuilt, Lee convinced Stout that a larger wing was essential. Stout said, "Never mind that wind tunnel stuff. I will give you the section coordinates." That Friday night, a Fokker flew in and parked on the other side of the field for the weekend. On Monday morning, Stout handed Lee the proverbial "back of the envelope" with a set of penciled coordinates. When Lee laid them out, he was quite impressed with the section. It was similar but somewhat thicker than the Clark Y airfoil used on the Fokker.

The Ford Tri-motor was built and was a sensation. John finally was able to test the section in the wind tunnel and discovered that its performance was superior to that of the Clark Y, probably because it was somewhat thicker. Later, John learned that on that fateful Saturday night, Stout and his henchmen had taken a length of copper tubing to the Fokker and carefully wrapped it around the wing when no one was there to see their skullduggery. They carefully brought the tubing back to Stout's hangar, laid the tubing gingerly on a sheet of paper, and traced the cross section. Despite their care, they must have stretched the section a bit and accidentally improved the performance.

I wish my memory were better because John was a *raconteur par excellence* with many wonderful stories about the early days. For example, there was a society meeting in Hartford at which John was the master of ceremonies and Otto Kopen, the MIT professor who among other things had invented the Aerocoupe single-control airplane, was the speaker. After the meeting ended, most of the audience remained to hear John and Otto reminisce about an early flying boat on Long Island. The story was hilarious; unfortunately, the details are gone from my recollection.

John Chamberlain, whom I shall introduce later, remembered one particular story told by Otto Kopen. It seems that one of the early aircraft designers would give detailed instructions to his test pilot before each flight to perform specific maneuvers, some of which were quite violent. The pilot noticed that after each flight, the designer walked slowly beside the airplane while mumbling to himself. After the third flight, the pilot asked the designer what he was doing. The designer replied that he was "counting the popped rivets." So much for computer simulation.

My first assignment was to calculate the temperature rise due to the combustion of hydrocarbon fuels over the complete fuel/air ratio range. With the careful tutelage of my boss, Dr. Neuman A. Hall, my engineering aide and I used a Marchant desk calculator (prior to the advent of computers) to solve the five simultaneous equations necessary for the calculations. After approximately six months of this dog work, Hall published a Society of Automotive Engineers (SAE) paper on the subject, listing me as co-author. The paper became the standard for the industry until the arrival of the computer. The paper also was significant at the time because of the development of afterburners and ramjets, which operated at high temperature levels.

Ramjets had fascinated me since my school days when, in Professor Edward Taylor's class at Massachusetts Institute of Technology, we learned about a jet engine for high speeds, which operated with no moving parts and achieved its compression solely through its forward velocity. The engine had been invented several years earlier by a Frenchman named Leduc and had been studied by the Germans, who called it the Aero-Thermo-Dynamic-Duct, or ATHODYD for short. The U.S. Navy had become interested in such an engine for a submarine-launched vehicle, the Rigel, proposed by Grumman. The vehicle needed a pair of ramjets of approximately 30 inches in diameter in order to cruise at Mach 2. Few air supplys in existence were adequate for the development of such a large ramjet, and we sold the Navy on a concept called the Multi-Unit Ramjet (MURAM), as shown in Fig. 1.2.

John Lee had the idea that a large engine could be assembled from a cluster of smaller units because the ramjet was so simple. We were to build an air supply in the Research Laboratory that could be used to develop the single unit. Seven of the units could be clustered to build the 30-inch size, which could be tested in the Willgoos Laboratory at Pratt & Whitney or at the Navy test facility, the Ordnance Aerophysics Laboratory (OAL), at the Lone Star Steel mill in Daingerfield, Texas.

To cluster the individual units, internal compression was essential, and the view from the front resembled a cookie-cutter shaped like a beehive. The units had a supersonic section, which began as a hexagon and contracted to a circle at the throat. The ratio of the capture area to the throat area, in accordance with the laws of physics, should be a certain number, as a function of the flight Mach number, in order to achieve optimum performance. As the flow decelerates in the contracting section, the Mach number ideally should go from flight level (in this case, Mach 2) to near Mach 1 at the throat. The starting requirement, in which the flow through the contraction is subsonic and total pressure drops across the external normal shock before it is swallowed, limits the contraction ratio to a much lower value. The

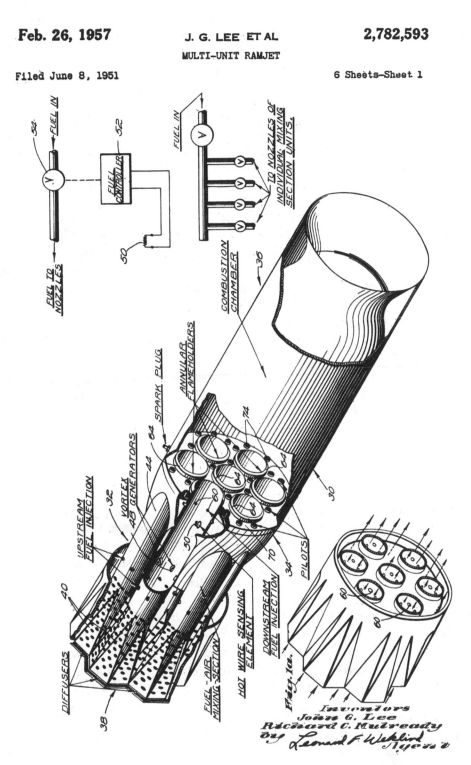

Feb. 26, 1957   J. G. LEE ET AL   2,782,593
                       MULTI-UNIT RAMJET

Filed June 8, 1951   6 Sheets-Sheet 1

Inventors
John G. Lee
Richard C. Mulready
by Leonard F. Weklind
Agent

Figure 1.2 Patent drawing of the Multi-Unit Ramjet (MURAM).

5

limiting contraction ratio had been the subject of a paper by Professor Arthur Kantrowitz and was known as the "Kantrowitz Contraction." By the strangest coincidence, his path and mine were destined to cross several years later in a different field.

To provide a way around the starting contraction limit, John Evard, an aerodynamicist at the National Advisory Committee for Aeronautics (NACA) Lewis, perforated the contracting section with a series of holes. In the subsonic regime, some of the incoming flow turned and passed out of the holes, which made the throat seem larger—large enough to swallow the shock. With supersonic flow established, most of the flow went shooting by the holes and was contained. This allowed a higher operating contraction ratio to be used.

George McLafferty, who did the supersonic development work on our program, decided to carry the process one step further by inventing the "Educated Hole," which resembles the hole in a kitchen cheese grater. This device further reduced the supersonic flow leakage out of the holes without blocking the subsonic flow. George was using a law in supersonic flow (Prandtl/Meyer), which states that supersonic flow can be turned only by a certain angle, even into a vacuum. Because no such limit exists with subsonic flow, George had improved the automatic shutoff function of the holes. In fact, he had to back off a bit from his initial attempt because it was saving too much of the low energy boundary layer. In the Research Laboratory wind tunnels, George tested a variety of single-unit diffuser configurations up to approximately one and one-half inches across the flats. He then clustered seven units, of the best configuration, and tested the multi-unit model in the Navy OAL tunnel in Daingerfield, Texas. George eventually achieved more than 88% pressure recovery—the ratio of the total pressure at the exit of the diffuser to the free stream total pressure)—for a single unit (Fig. 1.3), which was a record for its day for internal supersonic compression.

Here I must admit to a character flaw, which began to surface in those days: my abhorrence of the detailed rules promulgated by accountants, and a concerted effort on my part to work around those rules. George needed a change in the test section of the small supersonic wind tunnel in which he was doing the development work. The new test section was a duct, three inches by six inches by three feet long, containing the nozzle blocks, which determined the Mach number. There were glass windows on the side and a provision for holding the model under test. Mark Granville, one of the brothers of the "Gee Bee" aircraft fame, was foreman of the experimental shop and said he could "chew it out" in less than a week. The accountants had decided that an item called "test section" must be capitalized. Time and high-level approvals would have been required to place the item into the capital budget. Therefore, we called it a "model support," which was an acceptable charge, and thus we were able to expense the cost. The program proceeded without interruption and at a lower total cost than the wait would have accrued.

The Granville brothers, whose shop and hangar were located on the Springfield Airport in Massachusetts, built all kinds of unusual airplanes. The most famous of these was the "Gee Bee" racer, in which Jimmy Doolittle won the Thompson Trophy race. Mark Granville was a wonderful character and approximately an inch shorter than he had been as a young man. He had survived a crash during a test flight in a canard airplane called the "Ascender," built by

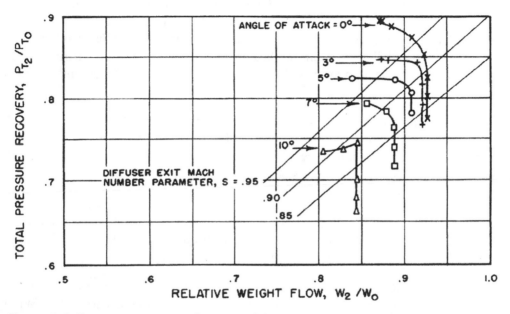

Figure 1.3 Pressure recovery, the ratio of the total pressure at the exit of the diffuser to the free stream total pressure. (Courtesy of Pratt & Whitney)

the Granville brothers. When Mark was fused together after the plane crash, he was an inch shorter.

One thing that evolved from the work on the Multi-Unit Ramjet was the fuel control. The search was for a simple, low-cost system, suitable for a throwaway missile. The hot wire, for measuring velocity in an air stream, has been used for some time. The heat transferred from the wire is a function of the pounds per second that pass. The clue was that it is necessary to correct for density to achieve the velocity. Properly set up in a bridge circuit, the voltage across the bridge measures airflow directly. The plan was to use this voltage to control fuel flow.

While George was working on the front end, another genius named John Chamberlain was working on the combustion problems at the back end. John has an innate feel for combustion and fluid flow, and he can answer a complex question involving those disciplines "off the top

7

of his head"—that is, within 5% of that produced by a team working with a computer for several weeks. John's early struggles provided the basis for designing high-velocity burners, which were both stable and efficient. NACA had identified "screech" (the high-frequency oscillation that seemed endemic to the afterburners and ramjets of those days) and later rockets (to be associated with radial oscillations) that could be effectively reduced with acoustic damping. John and his group, both at the Research Laboratory and later at Pratt & Whitney, developed a practical system to apply this theory, and Pratt & Whitney burners became free of oscillations.

In addition to his basic combustion work, John began to investigate ways to build efficient facilities in which to test combustion systems. This work had a major impact on all of Pratt & Whitney's future facilities. John recalls early tests of a two-inch ramjet that was firing behind the combustion test building. As John walked around the building while wearing ear muffs and carrying a sound meter that measured 120 decibels at 60 feet, he felt nervous every step of the way.

One of John's first major facilities was the jet burner test stand, in which he provided the concept and the construction follow-up. To avoid the need to obtain electrical drives that were in short supply in that critical period after World War II, John used war surplus Packard marine engines to drive the compressors, and these worked quite effectively. The airflow was sized to the 11-inch single-unit ramjet.

With a suitable test facility in hand, John developed a single unit of the MURAM with a slotted annular flame holder with individual pilots for each unit. They were 11 inches in diameter, and a cluster of seven would compose the 30-inch MURAM.

As the initial development of the single unit proceeded, it became apparent that the program had passed beyond the research stage and the project was ready to move to Pratt & Whitney. It was necessary to do the complete, full-scale flight weight design. Shortly, the cluster of seven units would be tested at the Navy Laboratory in Texas and at Pratt & Whitney's Willgoos Laboratory.

In September 1952, the Meteor (a twelve and one-half inch Navy airborne ramjet missile) and the Multi-Unit Ramjet projects were transferred to Pratt & Whitney, along with Don Brendahl and me as assistant project engineers for the respective projects. Fortunately, the brains such as John Chamberlain and many of our supporting test engineers also accompanied us.

From the relatively modest pace of the Research Laboratory, the change was to one of ordered frenzy. At that time, many new engines were under development at Pratt & Whitney. The company had been kept out of the emerging jet engine business during World War II because the government wanted Pratt & Whitney to concentrate on its piston engines, which were vital to the war effort. The corporation was spring-loaded to catch the competitors, who already were delivering centrifugal compressor-type jet engines. Pratt & Whitney made a gutsy investment in the Willgoos Laboratory (Fig. 1.4) to provide complete in-house test facilities for jet engines. The dollars involved represented a major part of the net worth of the company.

*Figure 1.4 The Willgoos altitude facility in East Hartford, Connecticut.
(Courtesy of Pratt & Whitney)*

The late start may have been a blessing in disguise for Pratt & Whitney. Jet production began with a centrifugal compressor engine, the J42, based on the "Nene," licensed from Rolls-Royce to "tide over" Pratt & Whitney's manufacturing plant. The major development effort under a U.S. Air Force contract was on a new, dual rotor, axial flow compressor-type engine. It was more efficient than that of the competition because it had a much higher pressure ratio. Known as the J57 (JT3), the engine stole the march and for years was dominant in various versions, both military and commercial aircraft.

Pratt & Whitney had evolved a strong project engineering system, which was unique in the industry. As in many companies that have operated for a long time, powerful bureaucracies had developed in the operating functions at Pratt & Whitney. Often, these groups did not see eye to eye, with much finger pointing and a general unwillingness to work together.

In late 1939, Luke Hobbs (Fig. 1.5), the engineering manager of the parent company, United Aircraft, wrote a historic three-page memorandum that defined the responsibility and authority of the project engineer. (See Appendix A.) Except for financial matters which were specifically excluded, the memorandum defined all responsibilities, starting with the initial design of an engine to its acceptance by the customer in the field. The memorandum put the authority for all aspects of a project in one place. In one simple line, Hobbs said that the functional departments were required only to facilitate the work of the project engineer.

Figure 1.5 Luke Hobbs was corporate engineering manager of United Aircraft in 1939. (Courtesy of Pratt & Whitney)

The system gave the boss a single button to push for each project. All efforts focused on making the product the best it could be, and any problems that arose were, for the most part, solved with dispatch. The environment for the project engineer and his assistants, inevitably dubbed APEs, was nirvana. The project engineer's only target was to develop the best possible engine in the minimum amount of time. He was supported by the most expert help in each discipline and faced minimal interference. The system seemed particularly applicable to the highly classified programs and, in many respects, anticipated the "Skunk Works" approach.

The project engineering system does not exist today, at least in the engine business. Engineering is no longer the driver. For the most part, technology has progressed, and although engines will become larger and control systems will improve, little remains in the way of component technology. Thus, performance cannot improve significantly without some change in mode. In those days, many new avenues were open, and basic component efficiencies were in the high 80s. Now, those efficiencies approach the high 90s. In addition, development cost was modest compared with the dollars produced by the high volume of production business. A development contract with the government might consist of three pages, stating that it was Pratt & Whitney's intention to develop an engine with certain power and weight and typically would cover approximately one-third of the cost to do the job. The balance of the cost was recovered in production overhead. You can understand why bean counters would be nervous with such limited control. The lack of project engineers having any financial responsibility contributed to the eventual downfall of the system, but it was great fun while it lasted. The inordinate growth of government penetration, where its local teams often measured in the hundreds, and the need for them to second guess all decisions meant the end. Today, it seems to take much longer to accomplish anything, and I bet things cost more now. The strong project engineering system was particularly applicable to the advanced emerging technologies, which often required a radical approach.

As I finished writing this paragraph, I discovered an article in *Forbes* regarding the extremely successful operation at Hewlett-Packard. I realized that, with one major exception, the project engineering system was not dead. It had simply moved to a field with rapidly emerging technology. In a highly competitive industry where the right technical decision is the difference between success or failure and with strict financial goals, these "nerdy" and streamlined "project" cells who had almost complete independence were growing rapidly and outdistancing the more conventional MBA-led competitors. It makes your heart feel good.

I believe in luck, and I am one of the luckiest. I had the good fortune to be assigned to two men at Pratt & Whitney who approached life differently but were both men of superior character. My immediate boss was Dick Coar, newly named as project engineer for ramjets. I worked directly for Dick for most of my 37 years at United Aircraft/Pratt & Whitney on a variety of extraordinary projects, as discussed later in this book. Dick has one of the finest technical intellects I have ever encountered, and it is seasoned with a broad interest and understanding of various fields. His skills were recognized early by those in authority, and he rose first to be president of Pratt & Whitney and then later to be a senior vice president of the corporation. In 1999, Dick was awarded the Guggenheim Medal for his outstanding achievements in the field of aerospace. Figure 1.6 is the only picture of Dick Coar that I could find. It was taken in 1981, on the occasion of my thirty-fifth anniversary when he was president of the Pratt & Whitney group. We were both younger and thinner then.

The other man to whom I was assigned was Bill Gorton. As development engineer, Bill was Dick's boss and had a reputation for being tough but fair. Bill also was on the fast track at

Figure 1.6 Dick Coar (right) and I (left) shake hands at my thirty-fifth anniversary at Pratt & Whitney. (Courtesy of Pratt & Whitney)

Pratt & Whitney. For all Bill's toughness, I learned over the years that he had a tender heart and was always ready to help if you were in trouble. Bill was our boss until Dick and I went to Florida, and later he came to Florida to become general manager of the Florida Research and Development Center (FRDC). Figure 1.7 is the only picture of Bill that I could find. It was taken after I had retired, when we had gone sailing off Palm Beach.

Figure 1.7 Bill Gorton (left) on a sailing excursion with me. Alden Smith is shown in the center, with my daughter-in-law Sarah on the right.

No schooling existed to teach you the Pratt & Whitney way. An APE was supposed to have learned by doing as a test engineer. It was not written—you simply were thrown into the system, and it was either sink or swim. When I arrived on the scene as an APE, never having been a test engineer, I survived only through the grace of my bosses and the outstanding test engineer, George Zewski, who had been assigned to me. The only lever I had was the four years of working on ramjets, with which they were not familiar. That benefit lasted about half a week with Dick Coar and a bit longer with the others.

One of the best sources of learning was Bill Gorton's Saturday morning meeting. The messages were primarily subliminal and were picked up almost by accident. One of Bill's messages that lasted through my whole career was the following: Listen to the experts around you...Don't substitute what you know for their advice...unless they don't agree. Then pick one path and go as fast as you can...It's quicker to find out that it's wrong and try some other than to dither at the crossroads.

My long working relationship with Dick Coar was unique. As with many brilliant people, Dick had a relatively short fuse, and he always "called a spade a spade." He had absolutely no time for any type of bureaucrat, and he had a healthy understanding of how little we knew about any of the "far out" things we attempted to do. Dick and I communicated easily, and as the years and projects passed, we could almost read each other's thoughts. For my part, Dick created a wonderful environment in which to work.

Our first office at Pratt & Whitney for both the Meteor and the MURAM projects was a "Hanging Gardens," approximately 100 feet on a side, above the Experimental Sheet Metal Shop. At first glance and particularly if you know the sound of a big sheet metal sheer in action, our office did not appear to be the best place for a budding engineer. However, it turned out to be an excellent spot and was an important early lesson on the benefit of rapid communication. Bill Gorton's office was in one corner and Dick Coar's was in the other corner, and the source of most of our ramjet parts were found at the foot of the stairs.

The Experimental Shop at Pratt & Whitney had the capability of turning out modest quantities of any conceivable thing for experimental test and was set up to support the project engineers. Joe Ballard, a friendly and capable individual, ran the shop. When Joe retired, he was succeeded by Ed Granville, brother of Mark Granville and one of the Granville brothers and another outstanding man. Ed was of modest height and always wore a jacket. He was dapper in contrast to Mark, who liked to appear as if he were a country boy. Ed and my group were involved in trying to fabricate some technically complex things over the years. On one particular trip returning from Florida, in the way of greeting, Ed reached up and grabbed both of my lapels. He then charged that his worry over the manufacture of a heat exchanger for the Suntan engine (which involved the vacuum brazing, with a gold/nickel alloy, of four and one-half miles of tubing) had contributed to his first heart attack. Of course, Ed was kidding and always enjoyed a challenge—the more difficult, the better. Similar to the other Granville brothers, who all had come from the village of Granville in the White Mountains of New Hampshire, Ed was inventive and at a young age had built one of the first snowmobiles, having put skis on the front and home-built half-tracks on the back of a Ford Model T truck.

My good luck continued with the people assigned to help me: Jack Franklin, process planner for the Sheet Metal Shop, and "Shorty" Harper, parts chaser. Among the many things Jack taught me was how to design sheet metal ducts and flanges so they could be welded together accurately without expensive production tooling. A more important lesson was that the best and most easily fabricated designs resulted from an early consultation between the designer who provided the required technical shape and the man who knew the process of manufacture.

Shorty was with me through many projects and was an old hand in what is more formally known as material control. He followed all of the parts we ordered when the designs were released, whether fabricated inside or outside, through inspection and into stores. Shorty had a knack of knowing if a supplier was having trouble and could place the right amount of pressure on the supplier. We missed few deadlines. If I had neglected some necessary connecting pieces, Shorty would fill in the order to ensure that when we were ready to test, everything was at hand. This sounds simple but was an invaluable help to a neophyte APE.

The full-scale design of the 30-inch diameter MURAM, based on input from the work in the Research Laboratory, was done by the Installation Design Group under Alden Smith (Fig. 1.7) at the airport. The reasoning was that ramjets resembled cowlings more than turbines. The preliminary layout of the complete seven-unit engine was finished in short order, and Alden and I proudly showed it to Dick Coar and Bill Gorton. Dick seemed properly impressed, but Bill was troubled. Bill finally said… "[Wright] Parkins will kill me if we release that big piece [the supersonic/subsonic diffuser section] as one weldment." Bill's solution was to include seven disconnect flanges (one in the middle of each subsonic diffuser), and the first design was shown that way. Later, I convinced Dick that the risks of separation were high with such joints, and we were able to dissuade Bill from using them. However, I had learned something about the way things worked at Pratt & Whitney.

Around this time, Parkins, the general manager and a legendary character in the industry, took me with him on a trip to Washington, D.C., to see the U.S. Navy. He was giving a presentation about ramjets to a group of admirals, who were seated in the front of the room while I stood at the back. In true Navy fashion, the rank decreased from the front of the room uniformly to the back, as knowledge of ramjets increased inversely. Parkins was holding forth in great style, as I noticed that some of the junior officers near me were beginning to fidget and look around as Parkins strayed farther off base. I was standing with my back to the wall, trying to give Parkins the "high sign" without the men next to me being able to see my actions. The admirals did not seem bothered by Parkins' medieval approach to supersonic flow. He caught my sign and smoothly changed subjects to something he did know, without anyone being embarrassed.

On a later occasion, with Perry Pratt and Dick Coar, we again were at the U.S. Navy in Washington, D.C. We had expected there would be informal technical discussions across a table. Instead, we were ushered into a small theater. A gentleman from Johns Hopkins named Dr. Badder, a ramjet expert from their laboratory, gave a presentation. Because he stuttered naturally, I think he gave his presentation at 100 miles per hour to ensure he would not omit anything. After Dr. Badder was seated, the admiral turned to Perry Pratt, and Perry said, "Mr. Mulready will give the presentation for us." With no advance warning or preparation, Dr. Badder was a difficult act for me to follow. The whole concept of a presentation before a group was new to Pratt & Whitney.

Some single units were built and checked out at the Research Laboratory, at the jet burner test stand. Both the single unit and the complete seven-unit RCJ2 were tested at the OAL in Daingerfield, Texas. The Lone Star Steel mill had a large air supply and exhaust system needed for the centrifugally cast pipe business. The Navy had piggybacked a large combustion test stand and a wind tunnel on this supply, which was used on a non-interference basis with the steel mill. This wind tunnel was used to measure the pressure recovery and drag of a 4.7-inch diameter model of MURAM at Mach 2.0. (See Fig. 1.3) Hugh Bowles managed this operation and later came to the FRDC during the development of that facility. We were impressed with the roofs of the test stands, which were designed to lift off easily as an explosion relief, and with the impressive collection of parts in a barrel that had been found at the edge of the site after various "test incidents," as they were called.

Figure 1.8 shows the MURAM engine. It produced the required 2000 pounds of thrust at Mach 2.0. After assembly, it was boxed and loaded on a Flying Tigers' freighter at our field in East Hartford. George Zewski flew to Daingerfield to be there when the shipment arrived. George called when the shipment was overdue, and we learned that Flying Tigers had mistakenly offloaded the box in Chicago. Bill Gorton was overheard "explaining" to them, on the telephone, the likelihood of future business from United Aircraft. George says that the box then arrived almost immediately and that for a lowly test engineer, he was very well treated by the Flying Tigers' staff.

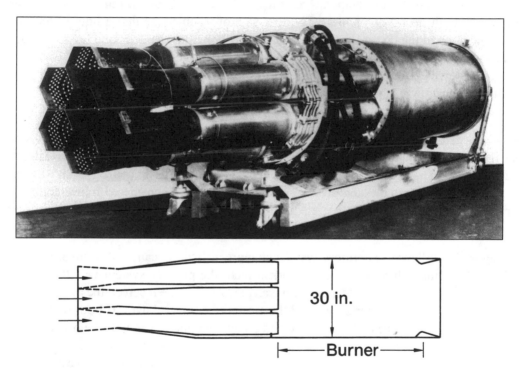

Figure 1.8 The RCJ2 Multi-Unit Ramjet (MURAM) engine. (Courtesy of Pratt & Whitney)

The skill of John Chamberlain with test stands was showing again. The vacuum pump was not large enough for the full flow of a complete free jet at Mach 2 for a 30-inch diameter ramjet. John had a bulkhead installed across the test stand approximately halfway down the engine, which separated the inlet spillover flow from the engine exhaust. The vacuum pump could handle the spillover flow; with a proper diffuser, the engine could pump its own exhaust back to atmospheric. Tests were performed up to an angle of attack of five degrees, which was quite an accomplishment. Pratt & Whitney provided the design, and OAL manufactured the Mach 2 free-jet nozzle with the understanding that it could use the nozzle for other projects.

During the first test, as the crew sat and peered through the porthole, water was filling the test cell and creeping upward toward the window. The test crew had forgotten that the water drain was manifolded both upstream and downstream of the bulkhead, and the cooling water was being sucked into the low-pressure region in front of the bulkhead. The water did not reach the porthole but came close to it. Water was splashing in the cell when Ray Kessler, who was studying the facility for ideas for East Hartford, turned to Dan Sims and said, "I always wondered why they called them battleship engines."

John Chamberlain remembers a more frightening problem with a heptane tank. The U.S. Navy required that the Meteor vehicle be tested with this fuel. A 400-gallon tank was parked outside the control room, approximately 20 feet away, and was connected to the facility. A steam heater controlled the temperature of the fuel in the tank, and it ran away. The pressure in the tank built up beyond limits. The first sign of trouble was that a liquid was creeping under the door to the control room. The test crew bolted to the door and opened it, and the crew was greeted with the sight of a geyser of heptane, almost as volatile and flammable as gasoline, blowing from the vent pipe on top of the tank. The safety blow-out patch had ruptured. The crew escaped as quickly as possible to wait for the pressure in the tank to drop. When the plume subsided, one of the test crew climbed up on the tank and screwed a cap on the pipe. There was no fire.

One day as the testing program was proceeding, George Zewski received a phone call to stop, pack, and return home. Bill Gorton had learned that, as a matter of course, OAL was sending all of our test results to the laboratory at Johns Hopkins. Bill considered Johns Hopkins to be our competitors. This meant that all future testing of the complete engine would be done "connected pipe," rather than "free jet," in our own facility at the Willgoos Laboratory.

Testing in the Willgoos Laboratory was much more difficult because we were approximately fifth on the priority list there. As it worked then, the priority system was that if priority number 1 was installed in his test cell and was ready to run, he received the air until he broke down. Then priority number 2 received the air until he broke down. If both were down and priority #3 was ready, he had his chance and could continue to run until he broke down. By this time, either priority number 1 or priority number 2 usually was ready, and the cycle would start again. The odds were slim that priority number 5 would ever have a chance. We finally convinced those in authority that an alternative system would give them what they wanted. The operating time in the facility total would be viewed as a pie and divided in slices, the size of which would be set according to the relative priority of the projects. For example, if priority number 1 had 60% of the pie, priority number 5 might have 5%. Each week, the available run time was divided, and each project was assigned a specific clock time for its air period. If you were priority number 5, you often received the third shift on Friday night; however, you can bet we were ready when our time came. The new system improved efficiency at all levels.

The Willgoos Laboratory was designed with four major test cells grouped around a large central bay in which the boilers from three Navy Cruisers and all of the turbines, pumps, vacuum pumps, and heat exchangers were located. This meant that all of the supply and exhaust ducts to the cells passed under the floors of the control rooms. With this arrangement, it was in the best interest of all concerned that a missed light, which occurs all too often with experimental equipment, could be quenched with water sprays and the fuel supply shut off as soon as possible. To this end, the fire insurance people were consulted, and an automatic fire detection system was installed, which employed a time-tested device called a "Fire Eye." This device was activated automatically by the test stand sequencer and would shut off the fuel supply five seconds after an attempted light, unless combustion was detected.

Alas, the Fire Eye detects the scintillation in a flame, and combustion in MURAM was so smooth that no scintillation was perceived and the system aborted every attempted light. All other data indicated that good lights were being obtained each time. Bill Gorton became more angry each time he heard the story, and he ordered a fix. The solution made use of the fact that with no moving parts, you can see from the front end through the combustion chamber to the exhaust nozzle of MURAM. The large air supply line passed under the floor, through the outside wall of the building, up through an elbow, back through a second elbow into the building, and into the test engine. A hole was cut into the top elbow in line with the engine axis and was fitted with a short pipe and an eyeglass peephole. An appropriate distance below, an iron farm tractor seat was welded to the pipe with foot rests welded below that. A test mechanic was stationed on the seat for each subsequent test and was given a hand switch to activate if he did not see stable combustion when a warning light came on five seconds after the start signal. We never missed a successful run after that. Lore has it that one of the earlier jet engine programs under Bill had problems with thrust meter repeatability. Bill fixed that problem with a weight equal to the rated thrust, a pulley, and some cable.

Several years later, after I had returned to East Hartford, I was shepherding a group of visiting Japanese technologists on a tour of the facility. As the bus passed between the building and the river, I was looking toward the back of the bus and saw two visitors talking to each other excitedly and gesticulating out the window. They had spotted the tractor seat, which was still welded to the pipe.

As we completed the successful test series, the U.S. Navy announced the cancellation of the program, having other fish to fry. There we were, with an operating 30-inch ramjet and no application. Boeing had a U.S. Air Force program at the time called the BOMARC, which had two 28-inch ramjets. We decided to give Boeing a pitch of our wares before we shut down. The BOMARC program was an anti-aircraft missile designed to go after enemy bombers. The major objection of Boeing to the MURAM was that it required solid rocket boost to near Mach 2 and that the empty booster cases might fall on the populace. In wartime as bombs were dropping, we thought a few empty booster cases was not a serious argument. We returned to East Hartford and closed the shop. Upon further reflection, it was concluded that the simple ramjet, while it required all of our expensive facilities to develop, did not have significant machinery content to be a good business for Pratt & Whitney. Within months, Boeing was experiencing a problem with its engine supplier Marquardt and came knocking on our door. We said, "Sorry. We are no longer in that business."

CHAPTER 2

T57—The Largest Turboprop

The next program for Dick Coar and me was the biggest turboprop ever built in this country, using the core of the successful J57 with a new low rotor and reduction gear. Dick Coar was the project engineer, and I was one of the APEs. George Zewski and Dick Adams, another old hand, were my test engineers. Bert Brown and Jack McDermott were the other APEs, both experienced hands from jet engine projects. Figure 2.1 compares the J57 with the new T57.

*Figure 2.1 Comparison of the J57 turbojet (bottom) and the T57 turboprop (top).
(Courtesy of Pratt & Whitney)*

The J57 is the military designation of the first dual-spool jet engine. With axial flow compressors that significantly increased cycle pressure ratio and efficiency over then current centrifugal compressor engines, it quickly became the choice for almost all military aircraft. The concept for the T57 was to take the successful high spool of the J57 and add a new low spool to build a turboprop. It required a four-stage low turbine, new low compressor, and a reduction gear to drive the prop. Four of these engines were to be used on the Douglas C-132 (Fig. 2.2), the first monster cargo plane, being developed for the U.S. Air Force. The aircraft was almost as big as a C-5 and had a maximum takeoff weight of approximately 500,000 pounds. I recall that the wing span was 178 feet.

Figure 2.2 Artist's conception of the C-132.
(Used with the permission of The Boeing Company)

Everything about this engine was enormous. It developed 15,000-shaft horsepower and had 5,000 pounds of jet thrust remaining. Four engines provided more than 170,000 pounds of thrust on takeoff, a huge number in those days. The new low turbine had the longest fourth-stage blade Pratt & Whitney had ever used, and it was the first time for us that a part span shroud was needed. A four-bladed Hamilton propeller, 20 feet in diameter with a blade chord of 2 feet, was required. It had separate bearings and an oil system, located in a housing that bolted to the front of the engine.

I was responsible for the development of the reduction gear to turn this big prop. To someone who had been dealing with no moving parts, it was eye opening. Al Rockwell was the designer, and between Al and my chief gearbox mechanic on the assembly floor, I was to receive in short order an education on reduction gears. With the advent of jets, the large size reduction gear knowledge, which had been highly developed at Pratt & Whitney, was falling into disuse. This program arrived in time to save it. Two of the repositories were Al and, as I recall, a man whose first name was Everett.

Al was what was known as a maverick in the U.S. Navy, having reached senior design engineer status without any formal engineering training. Al invented new concepts and designs that had an elegance unmatched worldwide. Everett had a natural feel for these big gears and could spot incipient trouble before it blossomed. He and Al had great respect for each another.

In the test rig, the two reduction gears are mounted nose to nose, connecting the input splines together with a long shaft, and connecting the output splines together with a short tube. Hydraulic cylinders are arranged to rotate one housing relative to the other, thereby loading the gear trains in both housings. The hydraulic pressure can set any desired tooth load. One set of gears is in effect loaded against the other. When the rig is rotated, only a modest input power is required to overcome the friction losses, while any desired power level can be simulated. We were able to test up to 17,500 horsepower with an 800 horse-power dynamometer.

The reduction gear was a two-stage planetary with individual planet gears that had a face width of two inches. The weight of each gear was approximately five pounds. Because at these high powers the teeth on the gears deflect in bending, Al provided both a tip and a root relief to avoid scuffing. The cantilevered planet support shafts from the cage deflect under load; therefore, Al skewed the teeth so they lay flat against the tooth on the neighboring gear at 80% power. Eighty percent was chosen because that is where most of the time is spent. The gear was tested over the complete operating range, and it completed all endurance requirements with flying colors.

George Zewski remembers a gag the floor mechanics would occasionally pull on a new engineer working with gears. They would cast a lifelike-looking gear out of "cerebend," the low-temperature alloy used for shop fixturing. They would cover the fake gear with grease to disguise it and then leave it on the bench. When the new engineer arrived on the scene, someone would ask him to dip the gear into the hot degreasing tank to clean it. They would wait for the horrified look on his face when he pulled up the empty basket. Then they would watch as the new engineer was handed a hook and began a frenzied thrashing in the bottom of the tank, searching for the lost gear until someone burst into laughter.

One of the real worries about such a big turboprop aircraft, particularly with the engines so far apart, was: What happens when an outboard engine loses power and the prop is not feathered? The answer is that it snaps off the tail. To reduce the likelihood of such an occurrence, the gearbox was fitted with an emergency negative torque control (ENTC), such that when a reversal of the load against the ring gear anti-rotation stop occurred, a signal was given immediately to the prop to feather. Such worries eventually contributed to the demise of the program.

One requirement imposed by the U.S. Air Force was that an external brake be fitted to one of the tower shafts. This would ensure that the props could be stopped from rotating on the ground due to the wind, rather than the Air Force having to cuff and sandbag the blades, as was the previous practice. Such a requirement was against our better judgment, but the brake box was designed and manufactured. Our concern was that the inadvertent actuation of the brake in flight would result in an almost immediate fire. Deciding that a demonstration was the best way to get the attention of the Air Force, a brake box was set up in a dynamometer stand. One of our best photographers, John Kulpa, rigged a high-speed movie camera on a boom to take photographs of what would happen. The rig was run up to cruising speed, and the brake was activated. The result was unbelievable fireworks as the magnesium gearbox burned. I cannot describe the painful look on John's face when he brought the camera into the control room. He had forgotten to remove the lens cover from the camera. We finally convinced him that because we were eyewitnesses to such a spectacular fire, we would be able to convince the customer without the photographs as proof.

Two new 33-foot houses were built for engine testing, which were the largest at Pratt & Whitney. Because the centerline of the engine was 16 feet off the floor, a means was necessary to facilitate service and repair of the engine. Thus, elaborate built-in gratings were needed, which could be moved quickly into place to provide access. One of my vivid memories is the day I stared out the control room window at Dick Coar and Bill Defienderfer from Hamilton Standard, who were standing on the grating in front of the engine. Dick was pointing to a large gash in the test house wall where the outboard half of a propeller had broken off and struck the wall at full power at the forty-ninth hour of a fifty-hour test. I had never seen Dick so angry, but Bill survived to become president at Hamilton Standard and later a corporate officer. During a dip in the stripping tank to repair the coating on a blade, a cleaning acid had seeped under the cuff, providing a "break here" perforation line, similar to the breaks in toilet paper rolls.

The engine also was tested in the big 20,000 horsepower dynamometer stand, #204, at the Willgoos Laboratory. After many attempts to make the new dynamometer work, I can remember the Friday night when an expert (I think his name was Bob Gorton) identified the problem. The monstrous magnetic field was lining up all the metallic chips inside the base castings that must have weighed at least 3,000 pounds apiece. It was creating a homopolar generator, and the solution was to remove the two pedestal bases, take them to the shop, acid clean them, and then repaint them. We did this work on Saturday, and we reinstalled the bases that night and ran the test on Monday. It was great to have the Pratt & Whitney shop next door when you were in a hurry.

With the tremendous flexibility at the Willgoos Laboratory of being able to operate at almost any flight speed and altitude, we quickly were able to map the complete operating range. I cannot remember the numbers; however, at altitude, it was better than a diesel.

The four major test cells, which each were approximately 75 feet wide by 200 feet long, were laid out in an "H" pattern with Cell 203 next to Cell 204. There was a J75 jet engine mounted next door in Cell 203, and the operator was named "Red" Tenant. One night, as we were starting up, Red came running through the door, waving his arms and yelling for us to stop. Our operator shut us down, and we ran next door to see what the trouble was. The engine-mounting plane was approximately 30 feet below the control room window, and the J75 had two large bleed valves located one on each side, 45 degrees from the top. The hilarious sight that greeted us were six-inch diameter streams of water arching upward from the bleed valves on each side of the engine. No wonder Red was excited! The exhaust system three-way valve, which normally would isolate the two stands, had stuck open, allowing our cooling water, which was a gusher, to leak over and fill up his engine. The engine dried and the program continued, with no discernible after-effects—except maybe for Red's laundry.

On one occasion, we received word that Wright Parkins was going to bring an admiral through our area on tour the next day to see the big turboprop run in the 33-foot house. The news made us panic because the propeller scavenge pump inside the housing had failed and there was no time to pull the prop for Hamilton Standard to repair. We decided to install a floor-mounted test stand pump to do the scavenge function, which the operator would turn on as we accelerated because the propeller system precluded turning it on in advance. The second shift operator was a great guy, dedicated to the program, and he remained there until early morning to have the pump installed. It worked fine on the test run.

Shortly after the second shift started, Parkins arrived with the admiral. We were gratified that our operator friend, who had come in for his shift, would have an opportunity to show off his handiwork. He started the engine and, after a pregnant pause, moved the throttle up to full thrust. Probably as a result of his lack of sleep, the operator forgot to turn on the scavenge pump. The excess oil caused the prop to go to full feather position. The loading down of the low rotor caused an immediate stall, followed by a surge in the system. The loud bangs and the big fireballs that shot from the front end were startling. Parkins and the admiral were bent down for a good view out the window. To his great credit, Parkins put his arm around the shaken admiral and said, "We run 'em hot here." They continued the tour.

On a previous smaller turboprop program, the T34, a B17 had been bailed to Pratt & Whitney by the Air Force to use as a flying test bed. The T34 was mounted in the nose of the airplane, and I can remember coming out of the Research Laboratory as it came over, on a low pass, with all four of the piston engines feathered. For the T57, almost twice as large, a C-124 was needed (Fig. 2.3). The nose door was locked closed, and a doubler skin was added to support the test engine.

Lead pigs were fastened to the aft ramp to counterbalance the weight in the bow. The balance was achieved, but the pitching moment of inertia increased significantly. The pilot remarked

Figure 2.3 The C-124 flight with test T57 engine. (Courtesy of Pratt & Whitney)

that it was similar to flying a dirigible. We calculated that the T57 alone would redline the airplane in a climb to 20,000 feet, but the Air Force made us fly with the piston engines idling. Don Riccardi, who was newly arriving at Pratt & Whitney, remembers seeing the C-124 on a test flight on his first day in the area.

Over the years, I have learned that disaster always seems to bring out the sense of humor in George Zewski. George and I worked together on many projects, and I learned that when I hear a little "heh, heh" from George, "the fit has hit the shan somewhere."

When testing in the 33-foot house, a large aluminum bell-mouth guided the air into the engine on the front. A heavy screen, supported by a rugged steel cage, kept out leaves, bugs, and birds in the front. We were running a series of surge tests with inlet distortion to determine the effect of varying degrees of distortion on engine stability. The compressor gurus had defined various sheet metal figures that were wired to the outside of the screen to produce the distortion. George called on Friday night. It seems that the screen, its frame, the aluminum bell-mouth, the angle iron, and the 3/8-inch steel bolts had all been sucked into the engine, with a great shower of sparks from the back end. The engine was so rugged with solid steel compressor blades that the damage was contained within the engine cases. The project was winding down, and we decided to file out the biggest nicks in the early stages that could be reached and finish the test. The damage was minor when the engine was taken apart. George Zewski remembers that all of the turbine blades had a handsome, uniform coating of aluminum, evidently from the melted bell-mouth.

The Air Force decided to move away from using turboprops and to build an all-jet fleet instead. Thus, we had another almost fully developed engine with no home. The few remaining engines eventually were used, without the reduction gear, to drive compressor rigs behind the Willgoos Laboratory. The low turbine was used with steam to drive compressor rigs at FRDC. So much for our largest turboprop.

Analytical studies in many places around the world had shown that a change to the turbofan, from the straight turbojet cycle, significantly improved performance. The fan works somewhat similarly to a propeller and improves the propulsive efficiency of the engine. With the J57 (JT3), a turbojet, Pratt & Whitney enjoyed a growing monopoly in both military and commercial applications. Wright Parkins, the engineering manager, did not want to hear the news that the fan was better. He personally told the world that it was not true. Chief Engineer Perry Pratt knew that the studies were correct, and he wanted some tests. Toward the end of the turboprop program, Perry told Dick Coar to cobble up a fan demonstrator from the JT3 and JT4. To avoid Parkins, we did the job in a satellite design office in Boston that had been taken over by Pratt & Whitney when the Chance Vought division moved to Dallas after the divestiture. These were the people who did not want to go to Texas. Our old friend, Alden Smith, managed the facility.

My job required me to fly to Boston each week to follow the design of the demonstrator while we were winding down the T57. The memory of one particular week in February 1956 remains vivid in my mind.

American Airlines operated the twin-engine Convair 240 from Boston to Hartford, and I was on the flight one Friday afternoon. It was a half-hour flight, and we had reached the Connecticut River when the right engine shut down and the propeller feathered. The prop had failed, and we returned to Boston. All the crash trucks awaited us in Boston, and we made a successful single-engine landing. After a short wait in the terminal, another airplane was pulled out, and we climbed aboard. The engines were started and then shut down almost immediately. The pilot had spotted a leak in the starboard engine, and thus the aircraft was returned to the terminal. A third airplane was pulled out, and except for some of the wiser folks who had departed, we climbed aboard again. This time, the pilot did not even start the engines because the copilot had spotted a leak during his walk around the aircraft. The wait was somewhat longer this time, but we eventually boarded the fourth plane and took off. As we approached the edge of the field, the starboard engine failed. The pilot made a second skillful forced landing amid the crash trucks.

By this time, the passenger list on our flight had shrunk considerably. However, after a half hour, we boarded for the fifth time and experienced an uneventful trip to Hartford in the second airplane, which now was repaired. In those days, American Airlines was run by C.R. Smith. Approximately two weeks later, I received a "bed bug" letter, under his stamped signature with a sentiment of "sorry for the inconvenience but safety is our most important product." I wrote a note in the margins, saying that this did not seem to cover four different airplanes and two forced landings for a half-hour flight. In two days, I received a personal letter from C.R., in which he agreed, explained all of the problems, and apologized. I was asked by many, including wife Carol, why I did not give up and return home by train after the first forced landing. My answer was that while I stayed at the Boston airport, I was only a half hour away from Hartford. Taking a taxi to Boston and then taking the evening train would have taken five hours. However, if you add the increments, it took five hours to reach Hartford by air anyway.

The demonstrator was built with a J57 core and the low spool from the J75. The layout for the J75 was spread out on a central table between the two designers. The designer of the "fan," the parts of which were taken from the low compressor, was on the board to the left. The designer of the turbine was on the board to the right. In those days, the practice was to sometimes put two different versions of an engine on the same drawing. Bill Gorton smelled something fishy and made us go through a laborious check layout to ensure that the dimensions were correct. Sure enough, a 1/4-inch error was found. It turned out that the designer on the fan end was copying from the version above the centerline, and the turbine designer was looking at the bottom version.

The engine was assembled and tested by other folks and proved the benefits expected from its modest bypass ratio. By that time, Dick and I had already gone to the next job.

Maynard Pennell, the Boeing 707 designer, eventually convinced our guys that American Airlines would use the General Electric aft fan if Pratt & Whitney did not do something in the way of a fan. Thus, after meeting with Pennell, Art Smith, Bill Gorton, and Perry Pratt sketched a fan engine in a hotel room in New York in order to waste no time. The first three JT3 compressor stages were replaced with two fan stages using JT4 transonic compressor aerodynamics. Two discharge nozzles were located up front, near the fan, and the package could be retrofitted on existing jet-powered airplanes. Pratt & Whitney moved quickly, and built and tested the engine almost overnight. It worked. American Airlines bought it, and the JT3D was born. The TF33, the military version, arrived a short time later, and the GE aft/fan was dead. This fan story is borrowed from the notes of a speech given by Dick Coar at the Elvie Smith lecture in Canada in 1989.

Because this is a short chapter, I have included two stories in it. They really have nothing to do with our projects, but they do etch some of the characters involved. They are about Wright Parkins and his cohorts, and they are worth repeating. Similar to all fables, errors probably exist. However, this is what I remember.

Wright Parkins was a strong, barrel-chested man, and he was rooming with Hugh Goslin at a business affair. Goslin was a wag of the first order of small stature, in contrast, but he was unmatched as a prankster, as you will see later. The party proceeded into the wee hours of the morning, and Goslin grew tired. He went up to the hotel room and went to bed. Sometime later, Parkins came up to the hotel room, no worse for wear; however, he had forgotten his key. Parkins banged on the door, and Goslin, who slept in the raw, answered it. As the door opened, Parkins peered in at this vision. Then he quickly reached in and grabbed Goslin by both shoulders, pulled the naked man into the hallway, jumped into the room, and slammed the door shut. Goslin pleaded with Parkins through the door, saying, "Please, Wright! Let me in! I can hear people coming."

Before you feel sorry for Hugh Goslin, continue reading here. Goslin, a true extrovert, was in the sales department and personally knew every airline president, admiral, or general who might buy an aircraft engine. At the time, he was stationed at the West Coast office in Los Angeles. Bill Gwinn, a relatively staid individual who later became chairman of the corporation, was manager of the West Coast office at that time.

Bill had bought a fancy new car and was taking Hugh for a ride down Wiltshire Boulevard. Goslin kept pestering Bill to see how fast the car would go, and Gwinn protested that he wanted to break it in gently. Goslin reached over, pulled out the hand throttle knob, and bent it downward. They were a wild bunch!

Dick Coar and I split for a short time, and he left to become the project engineer for the J75. In the morning before making the change, Dick was fighting to keep his T57 engineers from being transferred to the J75. In the afternoon, he was the raider. The J75 had had a major shaft failure as it was approaching production certification. Bill Gorton and Dick Coar were assigned to the project and determined that a major redesign was needed. Almost at the time the new design was ready for release, I returned to working with Dick, and we were off on our next adventure—liquid hydrogen.

CHAPTER 3

Liquid Hydrogen and the 304 Engine—Suntan

The next challenge was our greatest undertaking yet. It began for me when Dick Coar called me into his office one afternoon in April 1956 and asked what I knew about hydrogen. My smart-aleck answer was that it is a major part of water. Dick proceeded to explain that we had taken on the job of developing a special engine that used liquid hydrogen as fuel. It would fly Mach 2.5 at 100,000 feet, at a time when most other airplanes were poking along at less than half that speed. Only the U2 had approached altitudes of 80,000 feet. As was the case with all reconnaissance machines, the whole project was a secret, including the code name Suntan. The CL-400 (Fig. 3.1) was designed by Kelly Johnson at the Skunk Works. It was to

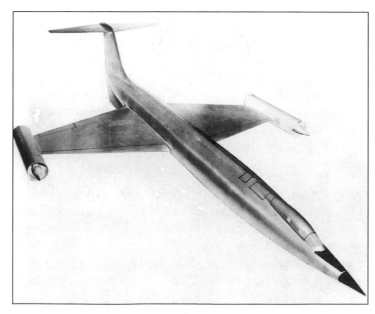

Figure 3.1 A wind tunnel model of the CL-400. (Courtesy of Lockheed Martin)

be a successor to the U2, which had become vulnerable to the Russians. The aircraft was to be 164 feet long and 10 feet in diameter, with a wing span of 77 feet. The shape of the F-104 fighter is unmistakable except for the two 304 engines, one on each wing tip.

This configuration would not have been successful. One of the major problems with the YF11, the predecessor to SR71 Blackbird, was the tremendous drag produced by popping the shock on one side while it remained swallowed on the other. The resultant torque tended to turn the airplane and could hardly be counteracted by the pilot, requiring rapid response to prevent losing control of the aircraft. With the engines at the wing tips, such correction would have been impossible.

Dick gave me the only two books we had on hydrogen at that time. The cover of one of the books showed a picture of the Hindenburg dirigible in red and in flames. The other book was actually a report of a fatal accident that had occurred in a laboratory in Toronto, where researchers had been working with a small Dewar of liquid hydrogen. These small flasks, resembling fancy Thermos® bottles, were dangerous when filled with SF-1, the top secret code name for liquid hydrogen. Soon after this time, we encountered a serious problem with such a flask.

This type of engine, called a hydrogen expander, was invented by a man named Rae and was of great interest to the U.S. Air Force for the reconnaissance mission. Theoretically, for an aluminum airplane, which is limited by the material melting temperature to a maximum speed of Mach 2.5, it would provide the highest altitude and best performance of any engine cycle and fuel. Because Rae did not have his own facilities, the Air Force asked for help from Pratt & Whitney. The Air Force approached Perry Pratt, chief engineer and unrelated to the original Pratt for whom the company was named, because Perry had been involved in the engine for the U2. Wesley Kuhrt at UARL was also approached. He later became chief scientist of United Aircraft. He invented the engine Pratt built, and it was much simpler than Rae's engine. Only a handful of people at United Aircraft were briefed on the project, and it was done on a strict "need to know" basis.

Liquid hydrogen has many unique properties, one of which is its extremely low temperature. At atmospheric pressure at sea level, it boils at –420°F. At this temperature, some materials such as conventional steels become brittle; others suffer from hydrogen embrittlement at high temperatures. Fortunately, some alloys are compatible with hydrogen, and the low temperature increases both strength and toughness.

This engine made use of the properties of hydrogen as a working fluid, with its very high specific heat and the highest heating value per pound of any fuel—almost three times that of jet fuel. For engines, hydrogen is the best of everything. It has a specific heat that is approximately 15 times greater than that of air. Thus, with a flow of only three and one-half pounds per second as the working fluid, it could produce the 12,000 horsepower needed to drive the fan by expanding hot hydrogen gas through the turbine, rather than using hot air as in a conventional jet. Figure 3.2 shows a schematic of the 304 engine.

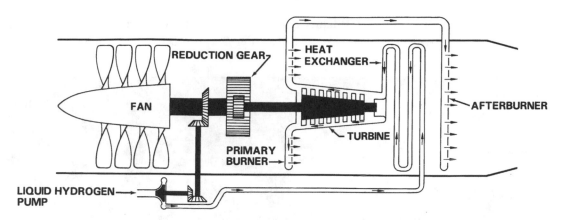

Figure 3.2 Schematic of the 304 engine. (Courtesy of Pratt & Whitney)

The major disadvantage of hydrogen is its low density—less than four and one-half pounds per cubic foot. A hydrogen tank with the same number of Btu's is five times larger than one for jet fuel, creating difficulty in designing a lightweight airplane.

To the uninitiated, the U2 jet engine resembled any other J57, and we could easily hide only those few things that would reveal its secret use. In the 304 engine, large fan blades were driven through a reduction gear by an eighteen-stage turbine, with some blades less than one-half inch long. Then there was the six-foot diameter, 72 million Btu/hour heat exchanger. Expediting the program meant using existing company facilities. "Green rooms," such as the final assembly area and the room where heat exchangers were assembled, were set up. These rooms were secure and were guarded twenty-four hours a day, because the most sensitive assemblies and functions were exposed. To enter through the lock required being on a special list and being personally recognized by the foreman in charge of that shift. Most of the detail parts were handled uncovered throughout the shop, with a tent covering only those assemblies and machine operations that were revealing. Wright Parkins was a great communicator, and, although he was not on the "need to know" list, he provided the other critical ingredient to keep the work secure. Parkins held a short meeting of all department heads. He told them that a secret program was underway at Pratt, and if he heard of speculation from anyone about what was occurring, that person would be dismissed. The assemblage left the room with their eyes straight ahead (similar to "blunk out," Lil' Abner's description of Little Orphan Annie). The feeling of terror apparently passed downward through all levels, and we never had a security problem. Everyone knew Parkins meant it. The other saving grace was the engine, which was unbelievable.

The first priority was to learn about hydrogen. Little useful information could be found in books, with almost nothing about liquid hydrogen. We were not privy to the large body of knowledge that had been accumulated for the hydrogen bomb work. The design started full blast in a green room, and a nearby safe place was needed where the material and machinery could be tested.

On the eastern side of the Pratt & Whitney airport in East Hartford, a heavily wooded section exists beside the runway. The "Klondike" test area (Fig. 3.3) was set up there almost overnight.

Figure 3.3 Aerial view of the Klondike test area. (Courtesy of Pratt & Whitney)

The test area had four component test stands for heat transfer, bearing and seals, and the liquid hydrogen pump with 1100 pounds per square inch discharge pressure. These stands were running in six weeks. On the engine stand, a converted J57 was tested on hydrogen gas from tube trailers in two months.

The white-roofed building, shown in the bottom left corner of Fig. 3.3, housed a modest liquid hydrogen plant, a cryostat that had been obtained from Professor Herrick L. Johnson of Ohio State University. He was a cryogenics expert who had built this cryostat for the U.S. Air Force in the early days of the hydrogen bomb program, when the bomb was "wet." Originally, it was thought that the bomb needed liquid hydrogen. When the bomb technology switched to hydrides and became solid, the professor obtained the surplus unit at a low price. When Suntan needed a source of SF-1 in East Hartford, the Air Force bought back the cryostat again for installation in the Klondike. Figure 3.4 shows an interior view of the hydrogen liquefaction plant.

Figure 3.4 The hydrogen liquefaction plant.

It was a turnkey deal, including a crew who taught our people how to run the system. In three and one-half months, the cryostat was producing 500 pounds of liquid hydrogen per day. Major Jay Brill, who handled all of the SF-1 logistics for the Air Force, occasionally would grumble about having bought the plant twice, but it served the program well. Johnson also ran a seminar to teach us the rudiments of liquid hydrogen.

To start heat transfer tests with liquid hydrogen before our cryostat was operational, Sy Bellak took a panel truck to Boston, where A.D. Little had a small cryostat, and brought back a modest Dewar flask of the liquid. Frank Williams remembers riding "shotgun" in a car behind the panel truck. Sy rode in the back of the truck for the return trip, and he had a small hand-held sniffer, with which he would regularly check the neck of the flask to make sure it continued to vent gas. Sy's face showed his concern as he drove into the Klondike. On Route 15 (now I-84) near the border of Massachusetts and Connecticut, there is a deep valley. Speaking *sotto* to avoid alarm, Sy said, "The normal venting did not resume after we climbed out of the valley." The increase in barometric pressure going down the hill evidently had forced something, most likely water vapor, to form a plug in the neck of the flask. This meant that pressure had been building up inside the flask since the vehicle had crossed the Connecticut line. The crew was evacuated, and the flask was placed on a bench that had been prepared. A gaseous helium supply had been previously teed into the transfer lance for purging. With Sy on one side and me on the other, we slowly lowered the lance into the neck of the flask. Whatever the plug was, it was melted by the warm helium, and the pressure was relieved. Sy went on to run our first heat transfer tests with liquid hydrogen.

The Air Force had 32 2000-psig tube trailers in inventory for transporting hydrogen gas over the road. We needed a supply of gas for the liquid plant and for other testing. A large hydrogen generator at a SPRY (trade name for saturated cooking fat) plant in Cambridge, Massachusetts, was used for saturating the fat. Twenty-eight of these trailers were shuttling back and forth between Cambridge and East Hartford, carrying hydrogen gas. Nine of them can be seen in the center left of the photo of the Klondike test area shown in Fig. 3.3.

Answers to most of the practical logistic and safety problems had to be worked out locally. One of the earliest worries was the flashback down the discharge pipe of a test rig during starting and stopping transients. The first solution was to bubble the discharge through a sand filter bed to act as a burn pit. It worked but proved to be unnecessary. It was important to ensure that any vented hydrogen gas was dissipated or consumed; therefore, a constantly burning pilot was fitted to the top of every stationary vent pipe. Figure 3.5 shows a burn pit.

Early in the project, it had been decided to modify a J57 jet engine to operate on liquid hydrogen and to provide an early flight test engine for the airplane before the 304 could be

Figure 3.5 Burn pit. (Courtesy of Pratt & Whitney)

ready. The engine was called "Shamrock," and the schematic in Fig. 3.6 shows how simple it was. Dick Anschutz was the project engineer.

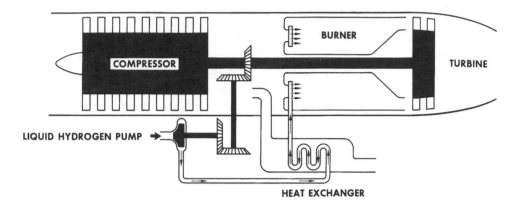

Figure 3.6 Schematic of the J57 converted for liquid hydrogen.
(Courtesy of Pratt & Whitney)

The modification of the burner dome required replacement of the complex liquid fuel spray nozzles with 1/4-inch tubes for hydrogen gas at each location. It worked correctly from the outset and produced a flatter temperature profile at the turbine inlet than the J57 had ever had. Figure 3.7 shows the burner can for the J57.

Figure 3.7 Burner can for the J57. (Courtesy of Pratt & Whitney)

The wide combustion limits of hydrogen permitted engine operation down to a temperature rise of only 200 degrees. The first-stage compressor blades could be seen moving slowly at this low speed. The acceleration to full thrust from this setting was very smooth, and operation on liquid hydrogen was better than it had ever been with jet fuel.

The pumps and the heat exchanger that ensured gaseous flow to the burners were tested as components. Dick Anschutz, who had come from the control group, used an early "breadboard" electronic fuel control on Shamrock and tested the engine on liquid hydrogen in the fall of 1956. The control truly was a breadboard, with the vacuum tubes exposed, resting on a bench and connected to the engine with wires.

Don Riccardi was present and remembers an early test. After the engine completed its run and was being shut down, the fuel control stopped responding to the operator's throttle lever. Don reached over and turned a knob on the control box, and the engine decelerated to a shutdown. It had been tested on hydrogen gas for 55 hours for checkout and had accumulated 2.24 hours on pumped liquid.

Don, who had joined Pratt & Whitney in 1955, had noticed copper-plated tools being made in the shop and guessed that something was happening. He went to Don Cleary, who was "den mother" for the test engineers at the time, and expressed interest in getting into something new. Don Riccardi was assigned to Suntan, which eventually brought him to FRDC—all because of his curiosity about the anti-spark tools.

With the start of testing of the 304 liquid hydrogen pump at the Klondike around seven months after go-ahead, a much larger SF-1 supply was needed than could be provided by the 500-pounds-per-day cryostat. To support the development of the in-tank fuel boost pumps for the airplane, which were being developed by the Pesco Corporation, Jay Brill had a 1500-pounds-per-day plant constructed by Air Products Corporation in Painesville, Ohio. The plant was called "Baby Bear." The Cambridge Corporation built a road-transportable Dewar of 3500 gallons, which resembled a standard gasoline tank truck (Fig. 3.8). The truck shuttled back and forth between Painesville and East Hartford to supply the Klondike. As I went to lunch on the day that the first truckload was to arrive in East Hartford, I was startled to see the truck parked outside a restaurant on Main Street, with the hydrogen vapor steaming away from the vent. The driver, completely unconcerned, had stopped for a bite before delivering his load.

Because of the low density of the fuel, only a single axle was required for the full-size tanker; thus, the vehicle was built that way. The driver told a story about being challenged at a truck weigh-in point. The operator of the station said, "…the single axle is OK while it is unloaded [it was fully loaded at the time], but don't bring it back here when you fill it up. It won't pass our load requirements…" A dummy axle was put on for the next trip, and we were never questioned again. Jim Smith, superintendent of all technicians working on hydrogen during the early days in East Hartford and later in Florida, is shown in the center of Fig. 3.8, standing with his hands in his pockets. He had a cool head and was always on the spot in times of trouble.

Figure 3.8 Truck for transporting SF-1 from Painesville to East Hartford.
(Courtesy of Pratt & Whitney).

The liquid hydrogen pump was a critical part of the engine, and that technology later would lead to the RL10, the first hydrogen rocket engine.

The pump rotor (Fig. 3.9) had two centrifugal stages located back to back. Aluminum star wheels were chosen for stress considerations at high speed; however, radial blades tend to produce instability when the flow is throttled. With the tight schedule and many unknown factors, a conservative structural approach was taken, and we lived with an occasional hiccup. The conventional oil-lubricated bearings on the shaft gave rise to a different problem. The second-stage impeller inlet operated at −410°F, and the oil-lubricated bearing, less than 6 inches away, was maintained at 165°F by a calrod wrapped around the bearing compartment. This complication was avoided later on the RL10.

To drive the pump on test (Fig. 3.10) with non-explosion-proof motors, the stand was designed with the drive located inside a pressurized room. The black metal can surrounding the pump is a vacuum jacket to reduce the heating of the cryogenic liquid by the atmosphere. Earlier, Lou Emerson had tried to insulate the pump with a foamed-in-place plastic; however, it always cracked, and the jacket was a direct approach. The inlet line and the overboard vent line also are double walled and vacuum jacketed.

Figure 3.9 Pump rotor. (Courtesy of Pratt & Whitney)

Figure 3.10 Liquid hydrogen pump on test. (Courtesy of Pratt & Whitney)

In the original engine design, the pump was to be driven by a small turbine with hydrogen gas tapped from the main stream flow. Dick Coar was concerned that gas pockets might be present in the near-boiling liquid being delivered by the airplane, which could cause cavitation of the pump. Cavitation could lead to an instant destructive overspeed if the pump became unloaded. The design was changed to a mechanical drive (Fig. 3.11), with a tower shaft geared to the main rotor of the 304. To control the engine, the system required a variable speed drive that could handle up to 1000 horsepower over a wide RPM range. The problem was that no such drive existed. Therefore, Dick and the design group invented it.

Figure 3.11 The tower shaft, hydraulic loading pumps, and recovery jacket over the hydrogen pump. (Courtesy of Pratt & Whitney)

Envision a planetary gear, where the tower shaft, down from the engine proper, drives the cage holding the orbiting planet gears. The sun gear in the center of the planets is connected to the pump shaft, and the ring gear is free to rotate in a large internal bearing save for two oil pumps, which act as brakes. The discharge from the oil pumps passes through a throttle valve adjusted according to the desired speed setting. With the throttle valve open, the oil pumps and the ring gear rotate freely, and little power is transmitted to the hydrogen pump idle. As the throttle valve is closed down, increasing the load on the oil pumps, the ring gear slows and the sun gear connected to the hydrogen pump increases in speed. In the limit, as the throttle valve closes, the ring gear stops and the sun gear operates at full speed and power transmission. This was an elegant solution to a difficult problem and a measure of the tremendous capability of Pratt & Whitney to solve a problem in a short time when all of its resources were unleashed.

The schematic of the 304 cycle shows the heat exchanger as the next component downstream of the pump. It was, without a doubt, one of the most fantastic parts ever installed in a Pratt & Whitney engine. In contrast to most jet engines, in which heated air drives the turbine, in this case the three-and-one-half pounds per second fuel flow, heated indirectly, was the working fluid for the eighteen-stage turbine. At full power, the exchanger (Fig. 3.12) transferred 72 million Btu per hour, enough to heat approximately 700 houses. It was six feet in diameter and was formed from four-and one-half miles of 3/16-inch diameter tubing. The alloy chosen was Hasteloy R® because of its compatibility with hot hydrogen. There were 1,200 tubes in the exchanger, with eight passes to a bank. Each bank was formed into an involute shape to maintain uniform spacing and blockage across the air stream. The two manifolds, which required precision drilling of the 2,400 holes to receive the tubes, were fabricated from forged Waspaloy®, a proprietary Pratt & Whitney nickel turbine blade alloy, which previously had been used only for cast turbine blades.

Figure 3.12 Exchanger build. (Courtesy of Pratt & Whitney)

Avoiding cracks in the manifolds, which could occur sitting on the shelf in a semi-finished state, required the development of a special heat treat with an oil quench. Frank Williams, the senior test engineer responsible for the heat exchanger, says that it produced his early gray hairs. The heat exchanger was the same assembly that Ed Granville, the shop manager, claimed had caused his first heart attack.

A high-temperature gold/nickel brazing alloy was chosen for its strength to fabricate the heat exchanger, and a ring of the alloy was slipped over the ends of the tubes as they passed through the holes in the manifold donuts. Gold/nickel foil also was used to attach anti-vibration clips to the tube banks. The whole assembly then was welded into a large flat can for the furnace brazing cycle. The can was purged with hydrogen gas during the brazing cycle to avoid oxidation, a standard brazing practice. Figure 3.13 shows the completed subassembly, ready for welding into the can.

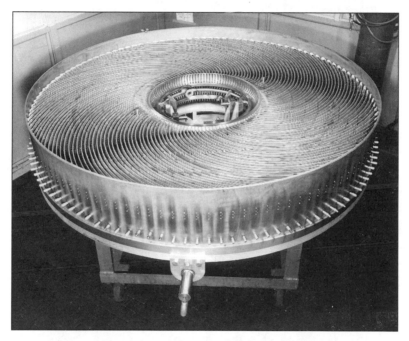

Figure 3.13 Braze assembly. (Courtesy of Pratt & Whitney)

Until this point, the assembly fabrication had been accomplished in a green room. With the exchanger covered by the can, it could be moved through the open shop to the furnace room without anyone noticing. Figure 3.14 shows the heat exchanger.

The same gold alloy was used in the turbine fabrication. At that time, gold had a controlled price of $34 per ounce. Each engine had approximately $25,000 worth of gold in it, even at this low price. In those days, the U.S. Treasury Department closely tracked everybody's purchase of gold. One day, we received an unannounced visit by a man from U.S. Treasury. He was of small stature but very determined. He wanted to know why Pratt & Whitney's recent purchase of gold had skyrocketed? The bean counters could not answer him and sent him to me. I was cordial and suggested that we take a tour of the shop to show him where gold was used. The shop has an area of 500,000 square feet, and we took him everywhere except, of course, into the green rooms. The tour took most of the morning, but the man obviously was not satisfied. He seemed to enjoy lunch but said as he left, "I'm coming back with a search warrant, and you will tell me where the gold is."

Figure 3.14 Heat exchanger.
(Courtesy of Pratt & Whitney)

It was time to grab the "red phone" to call the Air Force Special Projects Office for the Suntan program in Baltimore. They were a handful of people who were totally devoted to this project. The boss was Colonel Appold, and all of them had been selected with great care by the U.S. Air Force. Colonel Appold had been in charge of the Power Plant Laboratory at Wright field, which is the center of all U.S. Air Force propulsion work. During World War II, he led the bombing raid on the Ploesti oil fields.

I spoke to Major Al Gardner, who was in charge of the engine for the Air Force, and I explained the problem to him. With this kind of program, a direct connection to the White House existed, and we later heard that our Treasury visitor had been promoted and transferred to the Seattle office the next day.

The next component in the hydrogen flow path was the turbine, which resembled a miniature steam turbine rather than a gas turbine. It had eighteen stages, of which the first six were impulse and the last twelve were reaction. The straight, symmetrical impulse blades were machined integrally with the disk. The reaction stages, which were asymmetrical and twisted, were cast individually, and a saddle was machined into the root. The saddle was slipped over the edge of the disk, and each stage was gold/nickel brazed in a hydrogen atmosphere. Figure 3.15 shows the turbine, and Fig. 3.16 shows the eighteenth-stage blade.

Figure 3.15 Turbine. (Courtesy of Pratt & Whitney)

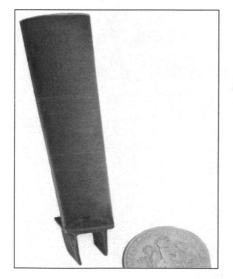

Figure 3.16 Eighteenth-stage blade.

Although thermodynamically a single part, the first six and the last twelve stages were mechanically divided and opposed to provide thrust balance. The two sections were connected with a spline and a shoulder in the center of the transition case. The hydrogen gas flowed forward through the first six stages and aft through the last twelve. The small diameter (approximately 1 foot) permitted the complete turbine to be located within the central body fairing. Figure 3.17 shows the turbine assembly.

The entrance to the turbine is located at the bottom. The last stage can be glimpsed through the colonnades in the transition case, dubbed the "Octopus." Maximum power was 12,000 horsepower at 25,000 rpm. The disks were extremely thin compared to any other Pratt & Whitney turbines and tended to form radial ripples after a few runs. It was something that had to be solved; otherwise, it was relatively free of trouble.

The materials were borrowed primarily from the jet engine. Problems arose because we were using high-temperature material, previously available in a cast form, as forgings. Heppenstahl, an old-time forging house in the southern part of Connecticut, took on the job of forging the turbine hubs of Waspaloy®, a tough material that had never been forged. The foreman called almost in tears. On the second blow, his drop hammer had shattered. The practice with most materials was to heat the ingot in the furnace, strike two or three blows, and then return the ingot to the furnace for reheat. We bought the foreman a new hammer, and our metallurgists suggested that he reheat after every blow. It was a slow process, but it worked and he was able to make forgings. However, the story does not end there. The machining of these tough alloys was difficult, and the Flannigan Brothers in Glastonbury, one of our loyal support shops,

Figure 3.17 Turbine assembly in central body.
(Courtesy of Pratt & Whitney)

took on the job. After experimenting with a variety of oils, cutting tools, speeds, and feeds, an acceptable surface finish finally was achieved. The machinist was making his final cut when the part broke into two sections in the machine—a perfect cup fracture. The reasoning was that the forging operation had produced powerful but symmetrical internal stresses in the part, compression in the core, and tension in the outer layers. The final cut removed enough of the tension material on the outside, which was holding the part together, and the internal compression overcame the remaining tension-holding fibers and caused the fracture. We returned to the heat treat and stress-relieving process and eventually produced good parts.

It may seem like a long time, but the turnaround time for recovery was rapid—often overnight. The test engineer had the authority to work with the shops directly, and together they found solutions in short order. Waiting to receive an approval from a boss was something we did not do.

The test engineers or APEs marked up drawings on the shop floor on the spot when they found a problem, and if the part worked, the marked-up drawing was recycled into the design. The process can become somewhat expensive, and more formality is needed later in the development cycle. However, with a devoted and knowledgeable team, the system cannot be beat in the early exploratory stage. In the ramjet days, Bill Gorton had taught us to make a decision and proceed. If the decision is wrong, take a different path and try again.

Because I always loved to "kick the tires," I acquired the habit of walking through the Experimental Shop first thing in the morning when I arrived at work. I knew where the "tall poles in the tent" were, and, one morning as I turned into the lathe section, I saw Dan Sims and Frank Williams huddled over a machine with the first six stages of turbine mounted. They looked at me and turned to one another and said as a chorus, "How does he know? He always shows up when there is trouble!" Dan, who was responsible for the turbine, had decided after a trial fit that he had to take another cut on the blade tips or it would never go together. It would take the part slightly below the drawing size, but Dan had judged that it was an impossible clearance, and it was his decision to make. It went together and worked, although the turbine designer always claimed that performance would have been better with a tighter clearance. That was a moot point if it would not go together.

Part of the discharge of the gas from the turbine passed into the primary burner in front of the heat exchanger. The flow was regulated to maintain the desired turbine inlet temperature, and the remainder was routed to the afterburner, downstream of the Hex. The combustion characteristics of hydrogen are so good that essentially ideal performance was obtained with an afterburner length of only half the diameter. The length of conventional jet engine afterburners is approximately five times the diameter.

The reduction gear was the next component in the mechanical chain. It was located in the central body, forward of the turbine. Its function was to transmit 12,000 horsepower while reducing the speed at the turbine from 25,000 rpm to the 6000 rpm required by the fan. This was my second project where Al Rockwell (mentioned in Chapter 2) was the gear designer, and again he produced a masterpiece, as shown in Fig. 3.18.

ONE FOOT   A-71

Figure 3.18 Reduction gear.
(Courtesy of Pratt & Whitney)

In contrast to Al's previous high-horsepower gears, this reduction gear was not a planetary gear but had six layshafts supported between two bulkheads. With the long flexible engine, alignment between the components was a major consideration. Al's solution was elegant. To split the load evenly between the layshafts, the sun gears floated axially between the pinion gears and were aligned by them. He split the sun gears in the center with opposing helical splines on their respective quill shafts. Close examination of the output sun gear shown in Fig. 3.18 reveals the split. To provide for a potential misalignment of the turbine and fan, the quill shafts had curvic teeth on both ends, which could accept considerable misalignment. Unless you love gears, this must seem a blur; however, it was a handsome package and weighed only 465 pounds, including the lubrication and scavenge system.

Next and last in the mechanical lineup was the fan, as shown in Fig. 3.19. It had five stages and handled 542 pounds per second airflow.

Figure 3.19 Sy Bellak, an APE on the Suntan project, stands beside the fan. Sy had been in my group since the Research Department days discussed in Chapter 1. (Courtesy of Pratt & Whitney)

The wide flight Mach number range of the vehicle—from 0 at takeoff to Mach 2.5 at altitude—meant that maximum pressure ratio was needed from the fan at takeoff and much less at maxium speed where forward velocity and the conical inlet provided most of the compression ratio. Variable stators were dictated to control their angles of attack. Linkages were connected to a crank on each airfoil. Because the variable stator was identified with brand

"X," Wright Parkins had forbidden its use at Pratt & Whitney. Fortunately, because he was not cleared, we never had to take Parkins into the final assembly green room. Don Riccardi remembers the first run of the fan in the Willgoos Laboratory—the performance was poor. After the run, Don discovered that all of the links between the actuator ring and the vanes were broken. A vibration during the start transient evidently had been the culprit. After the beef-up, they worked. There is not much to say about the fan because it did not cause much trouble. I am sure the guys in the trenches, such as Jim Ogan, remember the fan differently.

The first Suntan engine was assembled on August 18, 1957, and it was flown to Florida the next day by an Air Force C-124. Everyone was present on the assembly floor that night. The complex heat exchanger had lugs on the outside of the assembly, which were supposed to slide into channels riveted to the inside of the case. At midnight, we found an error where the channels were misclocked by ten degrees. The force against the face of the exchanger was significant; therefore, it was command decision time. Weld the lugs to the inside diameter of the case. It bent slightly when it ran, but it worked.

Dan Sims, who had been to FRDC, knew that at the time it was an undeveloped swamp, with none of the support we had been used to having. He recounted the euphoria exhibited by the foreman in charge of the test stands when he had found twelve stainless bolts in West Palm Beach. South Florida was not in a mode to support the test of the most sophisticated engine and fuel in the world. Dan had built a big mobile storage box, on casters, with shelves and drawers. When he told the tale of woe to Johnny Madden, who was the superintendent of assembly in Connecticut, Johnny took Dan and his support box through parts stores, and they pitched every conceivable thing they could think of that might be needed into the box. Dan struck his head, with an expletive, and said, "We forgot the ignition systems." Never mind. He would fix that. In a slouch, Dan is six feet, four inches tall, and he loped down to the Ignition Laboratory. A mesh fence surrounded it, at least ten feet tall. The fence was no problem for Dan; he climbed over it in a flash. Dan rummaged through the drawers until he found the model we used. It was the same as that employed on the jet engines. Dan did a "midnight requisition," literally, of a half-dozen units and climbed back over the fence. He could not leave a note because of the security, so we "winked." He put the six spares into the magic support box.

Some time later in Florida, we had an ignition problem, and we mounted one of the spares. Still, nothing worked, and this happened on all subsequent tries. Unfortunately, in his midnight raid, Dan had found the drawer for the "Defective—Return to Vendor" parts.

Figure 3.20 shows a later engine, a 304-2, which had a five-stage fan rather than a four-stage as on the first engine. It was mounted on a special road-transportable trailer and was covered with a stout wooden box. Thus completely covered in this way, it was towed to the airport. Here, the history becomes cloudy. Most of the people that I interviewed thought it was taken to our own airport, Rentschler Field, directly behind the plant. Dan Sims has the strong memory of towing the box to Bradley Field at midnight with a police escort, front and rear, with lights flashing. He says the Air Force was nervous about operating the C-124 out of our relatively short field and chose instead the 10,000-foot runway at Bradley Field for this first

Figure 3.20   The 304-2 Suntan engine.   (Courtesy of Pratt & Whitney)

flight. I believe Dan because it was his responsibility. The Air Force cinched up its courage, and all subsequent flights were made from Rentschler Field.

The drama continues. The flight proceeded to West Palm Beach, where Brandon Transfer, a local rigging company, had contracted to haul the box to the plant site. This was occurring before the Beeline highway existed, directly from the port of Palm Beach, northwest to the FRDC. The route at that time was up the Sunshine State Parkway to Jupiter and westward on Indiantown Road, out the back way. There was concern that the long and bouncy ride on a trailer with no springs could damage the engine bearings, and Brandon was strictly limited to a maximum speed of 15 mph. The only problem was that the minimum speed limit on the Sunshine State Parkway is 45 mph. A suspicious state policeman stopped our entourage, but we could not show him what was in the box. Dick Coar says I sweet-talked the policeman out of giving us a ticket for an oversize load, and we were allowed to proceed at 15 mph on the highway.

To provide security for the 7,000 acres of the triangular plant site, two of the sides were to be imbedded in the north side of the J.W. Corbett wildlife preserve. To obtain this land from the Florida Fish and Wildlife Commission, a third party had been hired to buy 9,000 acres from a ranch contiguous to the south side of the preserve, which then was ceded to the Commission in exchange for our site. All these shenanigans were done to hide the identity of United Aircraft as the purchaser to avoid the speculators.

Most of the roads through the site were in place, but the paving was incomplete. In typical Florida practice, a canal bordered the roads. These canals were called "borrow" pits because the material for the roadbeds had been scooped from them. It is called marl, and when the material becomes wet, it turns into slippery grease. That fall, we had 66 inches of rain, and thus standard practice at the end of a wet day was to line up behind the road scraper and creep from the test area seven miles back to the paved road. Dick Lindstrom, one of the TE's, loved to drift through turns on the grease. He went through a rental Studebaker about every two weeks. Figure 3.21 shows the B test area, looking northward.

Figure 3.21 The B test area. (Courtesy of Pratt & Whitney)

In Fig. 3.21, the B1 test stand is the structure on the left, and the B2 test stand is shown under construction on the right. The landfill for the stands was "borrowed" from the pit in the foreground.

Figure 3.22 shows an engine on the test stand. Figure 3.23 shows Sy Bellak inspecting the afterburner inside the engine. It was great to have a five-feet two-inches tall APE when a six-foot-diameter engine required inspection.

Here we were, a handful of engineers with the most sophisticated engine in the world, in the northern reaches of the Florida Everglades in an area called the Loxahatchee slough. The test facilities were still under construction, and the support crew was composed of all new hires. There had been a squabble about sending seasoned support personnel to Florida because East Hartford was very busy. In a huff, Wright Parkins said, "The hell with it. I'll hire them outside." Although most of the new hires tried, they simply did not know the Pratt & Whitney system, and it was not written down anywhere. Pratt & Whitney was a high-seniority organization, and many of the people there had worked together for more than a quarter century.

Figure 3.22 An engine on the test stand. The man wearing the checkered shirt is Al Gardner, the Air Force engine project engineer. Sy Bellak is barely visible, standing between the column and sprinkler pipe. (Courtesy of Pratt & Whitney)

Figure 3.23 Sy Bellak stands inside the engine. (Courtesy of Pratt & Whitney)

Those people knew how the company functioned and did not need a handbook. When a project engineer or his organization needed something, they were used to having whatever it was, in hand and on time after a phone call. None of the new crew, including the head of operations, Don Hazard, had worked this way in the past. Instead, they were used to bureaucratic systems in which things flowed through channels.

Soon, it became apparent that many things were falling through the cracks. A test engineer tracking a particularly critical item found that a facilities superintendent had filled a drawer in his desk with requests until he had enough on which to act. Our little group did an inordinate amount of follow-up, but things inevitably slowed.

There was one building in the C test area that we shared with the jet engine group by building a green room. All engine tear-downs and reassembly were done there, as well as instrumentation and "dressing" for tests. Figure 3.24 shows instrumentation in the C test area.

Figure 3.24 The C test area, instrumentation. (Courtesy of Pratt & Whitney)

Most people brought their own lunches to work, but an enterprising "Cracker," with "Sanitary Sam" painted on the side of his truck, would appear at the test area around noon to sell lunches. I never heard anyone becoming sick from his sandwiches and hot dogs, but it was always a possibility.

Both the electrical power and water lines to the stands were being installed. They came near one another in a trench behind the guard shack, and it seemed that on alternate days, the subcontractors who were doing the work would take turns digging up the other's lines. With all the problems, the engine, which had arrived in Florida on August 19, ran on the B1 test stand in September.

Lee Gaumer was the process engineer for "The Three Bears" liquid hydrogen plants constructed by Air Products Corporation for the Air Force. I remember him as a real go-getter who worked essentially as a project engineer would work at Pratt & Whitney. The first plant, "Baby Bear," which produced 1500 pounds per day, was located in Painesville, Ohio. This was the source for the liquid hydrogen that was trucked to East Hartford. With the engine development ready to begin at the FRDC in Florida, a much larger supply of SF-1 was needed. Two plants were to be constructed next to the Pratt & Whitney test area and were named "Apex Fertilizer Company" as a cover. Pratt & Whitney "loaned" a half section of land to the Air Force north of the B test area for the construction. We had a letter written by Al Flax, one of the few people in the chain who was cleared. He was Assistant Secretary of the Air Force for R&D, a well-liked and highly respected individual. The thrust of the letter was that the Air Force would return the land when it was finished using it. When the time arrived, years later it turned out that Al did not have the proper authority for such an agreement, and Pratt & Whitney had to repurchase the land from the Air Force. Ironically, a signature from a much lower level contracting officer would have been acceptable.

The first SF-1 plant built in Florida, called "Mama Bear," produced the liquid hydrogen for the early tests. It had an output of 7500 pounds per day from residual fuel oil. Figure 3.25 shows the facility.

The larger plant, called "Papa Bear," employed the new technology of turbo expanders to improve efficiency and throughput, and could manufacture 60,000 pounds of SF-1 per day from natural gas when everything was working properly. The new gas pipeline for south Florida had been installed the previous year and passed down the Beeline highway in front of the plant site. The plant went on-line in the spring of 1958 and served Pratt & Whitney as well as the early days of hydrogen usage at Cape Canaveral for the RL10. The view of Papa Bear shown in Fig. 3.26 was taken toward the south, and the B test area can be seen in the distance.

The output from "Mama Bear" was stored in four 28,000-gallon vacuum-jacketed tanks that were connected to the B test area with a double-walled, vacuum-jacketed pipeline (Fig. 3.27). The line was tested and worked, but it proved to be more efficient to move the SF-1 from the production plants to the test cells in road-transportable Dewars and run directly from these trucks. The relatively large quantity of fuel required to cool the line could be avoided this way. After the initial test, the line was not used during the Suntan program.

In early 1958, Pete Mans was working on the facilities. He went to Miami and rented a tank truck full of jet fuel so Lou Emerson could run a J57 on the B1 test stand. The facility officially was operational. It also gave the new and unseasoned crew a chance to run an engine before it had to run the liquid hydrogen engine.

Figure 3.25 Mama Bear, the first SF-1 plant built in Florida.
(Courtesy of Pratt & Whitney)

Figure 3.26 Papa Bear, the larger of the two SF-1 plants built in Florida.
(Courtesy of Pratt & Whitney)

Figure 3.27 The pipeline connecting the SF-1 storage tanks to the B test area.
(Courtesy of Pratt & Whitney)

For those who heard the 304 engine run, it was a sound they will never forget. It was dubbed the "Swamp Monster." The sound began as a low-frequency howl that would increase in pitch as the throttle was advanced until it became inaudible at full thrust. On one test, I held the phone out the test room door to enable John Chamberlain to hear the engine running. John was in Connecticut running the Combustion Group for Pratt & Whitney at the time. "What the hell is that?" was my question. After exhausting from the turbine, the hydrogen gas flowed out between two sheet metal bulkheads to provide uniform distribution across the burner. It passed into the combustion chamber through many small holes in the aft bulkhead, and the air passed through a forest of one-inch diameter tubes, which pierced both bulkheads and were arranged in an involute pattern to provide uniform distribution across the duct. The best guess was that the "organ pipes" were vibrating and being driven by high-frequency pressure oscillations in the combustion system.

After a few runs of the engine had been made, a visit was arranged for the "top brass." General Irvine, U.S. Air Force chief of staff, was hosted by Jack Horner, chairman of the board of United Aircraft; Perry Pratt, chief engineer of Pratt & Whitney; and my boss, Dick Coar. The control room was small and crowded. Therefore, it was decided to build a temporary observation point on the road that passed near the stand. Being outdoors would provide the advantage of being able to hear the Swamp Monster. A bunker was thrown up, and the post was connected to the operator in the control room with a headset. A check run of the engine was made early that morning to ensure that everything was ready for the demonstration.

As with all important visits, the order had gone out the previous day that everything was to be ship-shape and proper. A standard joke was that it was a good thing we occasionally had these visits because that was the only time we cleaned. Everything that could be painted was painted, and anything loose was tied down. The utility room located behind the control room contained all electrical and plumbing services. All instrumentation wires passed through there, neatly covered by sheet metal ducts. On second shift the night before the demonstration, a zealous junior electrician had noticed that if he shook an electrical conduit vigorously, it would rattle against a duct. He secured the errant conduit with a sheet metal screw through a clamp to the duct, as he had been taught in school. As luck would have it, the sheet metal screw had pinched the insulation on a wire inside the duct, a safety circuit designed to kill power in the event of a short.

As project engineer, I had to be on the headset when the visitors arrived. Having described to them what they were about to see and hear, the operator started counting down for the start. The next thing he said was, "We have lost all electrical power on the stand." Evidently, when the check run had been made in the morning, it was cooler and the screw had not yet squeezed through the insulation. As the Florida day grew warmer and time passed, the errant screw had created the short and killed the power at the time of the demonstration. Fortunately, the assemblage was most understanding. Unfortunately, none of them had the opportunity to hear the Swamp Monster.

In the winter of 1957–1958, Lockheed shipped a section of the airplane tank, boost pumps, and plumbing to FRDC for testing. Lockheed did not have test stands or a fuel supply available for the scale of testing required. The Pratt & Whitney test stand crew was fascinated to see this real, lightweight airplane stuff. With the help of the Lockheed people, the tank section was parked next to the B2 test stand in preparation for tests the following week. As nature would have it, the tests never occurred. On Sunday night, a freak tornado-like windstorm passed through the plant site. The storm caused relatively little damage, but it picked up that tank and carried it three-quarters of a mile into the swamp. It also picked up a Mac truck cab and bounced it several feet down the road while the driver ducked to the floor.

The engine test program went well. Five engines had been built for development, and the component work was proceeding. There were the usual problems expected in a new design but surprisingly few problems attributed to the liquid hydrogen fuel. For engines, SF-1 did everything well. However, this was not true for airplanes. The low density and the need for lightweight insulation made all of the aircraft problems difficult. Kelly Johnson was becoming more and more disenchanted with Suntan and SF-1.

As it happened, an excellent alternative was at hand. If you could fly faster than the aluminum limit of Mach 2.5 (say, Mach 3+), kerosene-type fuels came into their own again. The secrets were titanium and the convertible engine.

Both of these things originated at Pratt & Whitney and made the SR71 possible. You will not find any reference to either of these elements mentioned in the official Blackbird histories. In fact, it sounds as if all of the airplanes were high-speed gliders, with scant mention of the power plants that made the craft possible.

The potential to achieve high strength at high temperature with titanium metal was known for some time. The Air Force recognized that jet engines, particularly turbofans, could be significantly reduced in weight if ductile titanium metal could be produced. The Air Force contracted with Pratt & Whitney in the 1950s to pursue this course. From this major effort, things such as double-vacuum melting with consumable electrodes and welding in an inert atmosphere bag were developed. By the end of the 1950s, Pratt & Whitney was using more than 80% of the titanium metal produced in the United States. The Air Force-sponsored work at Pratt & Whitney encouraged a budding industry and developed practical manufacturing techniques for the metal in time to make possible the A11, and the A12 later known as the SR71.

The second major element was provided by Bob Abernethy's invention of the J58 turbo-ramjet. Bob was an analytical engineer, studying the performance of the J58 in the supersonic regime, when he realized that the compressor/turbine combination was producing a total pressure drop in the engine air flow at flight Mach numbers above 2.3. His idea of adding six large bypass pipes with valves from the inlet to the afterburner essentially turned the J58 into a ramjet above Mach 2. Flight at more than Mach 3 had been made possible, and Kelly Johnson was not one to miss such an opportunity. He returned the Suntan money and sold the titanium airplane.

This is a unique case in which two companies, one airplane, and one engine discover that the path on which they are proceeding is wrong and, not being in the spotlight, pursue an alternative path that becomes historic. The other interesting part is that the hydrogen technology was applied to a different field, and that has produced a legend, the RL10.

CHAPTER 4

RL10—My Only Moneymaker

The head of the Advanced Research Project Agency (ARPA), Roy Johnson, directed Norm Appold and the Air Force Special Projects Office to start a rocket engine program using the liquid hydrogen technology that had been developed for the 304 Suntan engine. It was to be used on the Convair upper-stage space vehicle later called Centaur. Dick Canwright, whose career had included stints at Jet Propulsion Laboratory and Douglas Aircraft Company, had come to ARPA and was given the job of working out the specific details and work statement with Pratt & Whitney. Krafft Ehricke and a few others at Convair were briefed on the work that had been going on at Pratt & Whitney. We were unaware that Convair already had a study underway with Rocketdyne for a hydrogen/oxygen stage that operated on pressurized propellant tanks. When Convair learned that we knew how to pump liquid hydrogen, it became excited.

A flurry of meetings occurred to define the engine, and it was decided that the 30,000 pounds of thrust required by the vehicle would be supplied by two 15K engines. The Air Force designation was LR115. We did not learn until much later that Krafft had chosen the 15K size because it could use the same pump size as the 304. Had we known, it would have been easy to double the size and build a 30K engine, which in most respects would have been easier for everyone. As it turned out, the RL10 pump was a new design, far superior to the original pump, incorporating all of the lessons learned.

One last question remained: Which oxidizer should be used? Fluorine offers a performance advantage over oxygen but is extremely reactive and more difficult to handle. Before the contract was finalized, we decided to run small-scale test firings to examine the performance and handling. Charlie King, who was our resident rocket guru, set up a firing position over one of the ponds behind a jet engine test stand. Fluorine gas was available in cylinders, but we needed liquid. Charlie set up the fluorine gas supply line so that it passed from the bottle through a trough flooded with liquid nitrogen on its way to the test chamber. The liquid nitrogen is cold enough to condense the fluorine gas, and thus liquid was delivered to the test chamber.

Fluorine is so reactive that it ignites spontaneously with anything that will burn and some things that usually do not burn. We learned much later that if fluorine encounters an already oxidized material, such as an ice cube, it would tear the hydrogen from the oxygen with an explosive reaction. The occasion was the introduction of liquid fluorine into a test stand line that had gone through an exhaustive purging procedure. As it turned out, the purging procedure was not quite exhaustive enough. Near an elbow, some water ice crystals must have remained in the pipe at the cold end. When the fluorine hit the ice, it blew the elbow off the Schedule 80 stainless pipe.

In the stand set up for the test firings, there was a pressure tap on the supply line submerged in the liquid nitrogen trough. The tap was connected to a pressure transducer approximately four feet above the trough with a one-quarter-inch stainless line. Evidently, a minuscule leak internal to the transducer had not been detected on the leak check with the soap bubble technique. The transducer caught fire, as did the stainless line. The stainless burned like a candle, being quenched only when the fire reached the liquid nitrogen surface in the trough. Somewhere I have a piece of stainless tube with little globules of metal adhered to the side, similar to droplets on a wax candle. It was the consensus that liquid oxygen (LOX) should be the oxidizer. At least, all the alligators in the pond had good teeth.

Years later, to his great credit, Bill Creslein successfully operated an RL10 on fluorine; however, it was always difficult, and the performance improvement was not worth the risk.

The original cost estimate for the development of the engine, through preliminary flight rating test, was $9 million. When it was time to sign the contract, Dick Coar called me into his office to discuss what number we should propose. This was to be the first cost-plus contract Pratt & Whitney had undertaken, and we were plowing new ground in all directions. As I remember the discussion, we both had our feet on the table, and Dick asked if I thought $9 million was enough. After some discussion about the things we had to do and the many unknowns involved, the estimate was doubled to $18 million, which was accepted. It turned out that the cost through the preliminary flight rating test of the A1 was $36 million.

The factor of two was the correct value; we simply should have applied it twice. Endeavors where all technical boundaries are being exceeded inevitably lead to cost overruns. You probably are not reaching far if you can claim being both on time and within budget. As it turned out, the program proved to be relatively straightforward, with only one nasty surprise.

Preliminary designs of the engine showed a conventional cycle, the gas generator, in which a small fraction of the propellants are burned in a separate combustion chamber to power the turbopumps. This cycle requires a separate propellant control and ignition system, among other things. Perry Pratt had shown some sketches to Al Donovan, president of Aerospace Corporation, the "think tank" for the U.S. Air Force. Al was an old friend of Perry's and knew about Suntan and the 304. He suggested we use the same cycle as the 304 and avoid the gas generator and all of its problems. When Perry returned home, the RL10 design was revamped, and its elegant simplicity became apparent (Fig. 4.1). At that time, Bob Atherton was working for Bill Sens, chief of all performance analysis for Pratt & Whitney in Connecticut. Bob and Bill had a major impact on the soundness of analytical design.

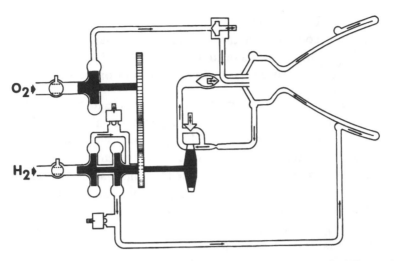

Figure 4.1   The RL10 cycle.   (Courtesy of Pratt & Whitney)

In this cycle, the waste heat rejected to the chamber walls provided the energy to drive the turbopumps. All of the extra systems and hardware of the gas generator were eliminated. The high turbine inlet temperature of the conventional cycle required special materials. With the expander cycle, the turbine inlet temperature was −165°F. The excellent properties of pure hydrogen as a working fluid could provide the 800 horsepower required at this low temperature. The turbine blades, stators, and cases all were made of aluminum.

It is interesting that of all the hydrogen engines developed later worldwide, no one selected this cycle. It must have been the "not invented here" syndrome or lack of courage. When the French needed a hydrogen stage for their Ariane, they first attempted to buy the RL10 for the job. Suddenly, the technology was too precious to the U.S. State Department because the Ariane might compete with the Space Shuttle and it could not be exported. It does compete, and it is unfortunate that we do not have a piece of the action.

The French decided to develop their own hydrogen engine. By that time, I was retired and their Washington representative approached me for a recommendation for a consultant. Without any questions, I recommended John Chamberlain, who also had retired by that time. John was invited to France to discuss the possibility, and some bright individual there decided to negotiate with John—not directly, but through a third party. After being forced to wait in a hotel room for some time, John grew weary of the delays and returned home. That has to be the worst negotiating tactic anyone has ever used.

Another excuse used to discredit our proposal for the 200K engine, discussed later in this chapter, was that the expander cycle could be applied only to small engines. This was grossly untrue, and in the early 1990s, a design was submitted for a 600,000-pound thrust split expander engine, with a turbine inlet approximately at room temperature. This design employed a thrust chamber fabricated from copper tubes. Later, Don Riccardi, one of the last of the original hydrogen "Mohicans," proved the critical manufacturing process. Don was notified that a patent for the copper tube chamber was issued in November 1998. The advantage of this type of cooling over that provided by the conventional milled channel chambers is a much longer life, with reduced thermal cycling damage in the difficult environment of a high chamber pressure rocket. Back to the era of modest chamber pressure.

It is hard to over-emphasize the significance of the cycle choice for the RL10. The singular physical properties of pure hydrogen, which the cycle exploits, allow very large margins of safety in design. The simple cycle completely eliminates many complex ancillary systems. This fact and the large design margins that are possible at low temperature lead to unmatched reliability. Starting is accomplished by opening the propellant supply valves and admitting hydrogen into the thrust chamber cooling tubes. The sensible heat residual in the tubes, even after many hours of coasting in space, is enough to spin up the turbine. The ability to restart in space offers the potential to exploit missions with multiple firings interspersed with periods of coasting. Over the past 40 years, the RL10 has been fired more than 550 times in space, with only three failures attributable to the engine. Some of the historic missions included the Surveyor lunar lander on the initial flights to the moon, the Viking Mars lander, the Voyager slingshot fly-by that now has departed the solar system, and myriad communication satellites. The first was the Comsat that introduced transcontinental television with the Olympic games from Europe. After the initial A1-1, models have included:

- RL10A-3-3A
- RL10A-4
- RL10-4-1
- RL10B-2

The last three models have deployable nozzle skirts to improve performance. The thrust ranged from 15,000 pounds to 24,750 pounds, and the specific impulse ranged from 420 to 466.5 seconds.

The third failure attributable to the RL10, of the B-2 model on its maiden flight, was caused by a structural failure of the thrust chamber. The failure occurred on the second firing in the flight after a coast period. It is reasoned that the rupture that occurred on the twelfth firing of the engine in its life was due to inadequate coverage in a brazed joint, which had not been detected by x-ray. Ultrasound has been added to the x-ray inspection of the chamber. An RL10B-2 was successfully launched on August 23, 2000. The B-2 description continues at the end of this chapter.

One final change from the 304 technology was the replacement of oil-lubricated bearings with those that are hydrogen gas cooled. The difficulties of conventional oil lubrication described in the previous chapter, where the necessity of maintaining liquid hydrogen temperature at the inducer inlet while keeping the bearings warm enough to avoid freezing the oil, produced 600° gradients along the first four inches of the shaft. To do this required a large calrod element wrapped around the bearing compartment, which was messy and made pump cool-down more difficult. I cannot recall who in the group first suggested that if the bearings were cooled, those bearings could be operated without oil. That person convinced me, and I tried the idea on Dick Coar when we were on a trip to San Diego. Dick said you can try it, but you must bring along the oil-lubricated box as a backup. Within three weeks of our return to the FRDC, test work to prove the concept had been completed, and oil lubrication was canceled. NACA Lewis evidently had worked previously on hydrogen-cooled bearings to eliminate oil; however, we were not aware of that effort. This change resulted in the ultimate design simplicity.

The engine Marque RL10 is already the longest lived at Pratt & Whitney still in production and is going strong after four decades. Almost four times as many RL10's have been produced than all of the engines used in Apollo. It is now a commercial product. The latest model is the RL10B-2 for the DELTA III. The original engine, the A1, had a rated thrust of 15,000 pounds with a nominal specific impulse of 420 seconds. Specific impulse is simply a measure of how many pounds of thrust you achieve for each pound per second of propellant burned. The B2 is rated at 24,750 pounds of thrust with a specific impulse of 466.5 seconds. Since the beginning, more than 450 engines have been produced. For comparison, the JT8D turbofan had a production run of more than 11,000 engines. Thus, rocket engines do not begin to challenge the jet business.

Over the years, competitors and opponents in the government have attempted to displace the RL10 without success. The first serious effort occurred in 1979 when the Space Shuttle, then under development, was going to provide all launch capability required. At that time, NASA took steps to shut down the RL10 and Centaur because it did not want anything to compete with the Space Shuttle. Intelsat, who at the time bought its launches from NASA, objected to the proposed cancellation of the Centaur because it knew the Space Shuttle was having serious problems. Intelsat complained so loudly that NASA was forced to continue the RL10 and Centaur. More discussion about attempts to kill the RL10 are described later in this chapter.

In the late 1950s, a massive struggle was occurring between the U.S. Air Force and NASA to see which organization would take the lead in space. An article, "From the Bay of Pigs to the Sea of Tranquility," was published in *Washington Monthly* magazine in September 1970, subsequent to the Apollo disaster. This article provides an in-depth look at the history of the early space effort and all of the "wheeling and dealing" that was happening at the time. The work was excerpted from a book written by three journalists (Young, Silcock, and Dunn) from the *London Sunday Times*. Doubleday published the book, *Journey to Tranquility*, in May 1971. The book provides an outsider's view of this fight and reveals how the deals made by President Lyndon B. Johnson, Robert Kerr, and Fred Black, lobbyist for North American, shaped the U.S. Space Program for decades. Because of their excellent politics and President Dwight Eisenhower's desire to put a peaceful face on space, NASA won the fight. Consequently, the RL10, which had started development under the U.S. Air Force, was transferred to NASA in 1961. The company designation RL10 was adopted by NASA.

The start of development of the LR115 (RL10), under Air Force contract, began in mid-October 1958. The development team was spring loaded to exploit this newly learned technology. Having recently put the 304 to bed, we had a cadre of engineers, technicians, and mechanics who had worked with liquid hydrogen for more than two and one-half years. With the elegantly simple design and no critical materials, we hit the ground running and accomplished the initial development in record time. Here I feel my inadequacy as a writer because

I do not have the skill to describe how extraordinary we all felt at that time. It was fun to come to work in those days.

Design of the engine was done in East Hartford, and some of the most talented Pratt & Whitney staff were assigned to the job. I cannot remember the names of all of the people involved, but everyone wanted to help. Two names that come to mind are Walt Ledwith and John Chamberlain. Walt moved to Florida and became second in command for the FRDC design group, under Ben Savin. Walt was a major contributor to the design of all future rocket and jet engines.

John was responsible for all combustion and heat transfer R&D in the advanced technology department for Pratt & Whitney under Walt Doll. John worked directly for Ed Brown, in the combustion area. Gene Szetela, in the heat transfer group, reported to John. Gene spoke softly in a quiet voice. However, when he talked about heat transfer, everyone listened attentively.

In this position, John's unmatched talents influenced all of Pratt & Whitney's projects and designs. Fluid dynamics and thermodynamics were second nature to John, and his innate understanding of how things interacted was a tremendous benefit to the company. John provided the combustion and heat transfer design input for the RL10. He also conceived of the unique test stand design (Fig. 4.2), which provided an excellent simulation of the space conditions under which the engine operates.

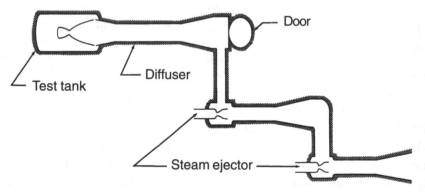

Figure 4.2   The RL10 test stand schematic.   (Courtesy of Pratt & Whitney)

The basic idea, which John had used in the early ramjet and turbojet facilities, is that the hot exhaust stream from any type of engine, when running, has enough energy to pump itself back to atmospheric pressure, provided that it passes through a reasonably efficient diffuser. The convergent–divergent diffuser took care of the steady-state requirement. To simulate the start in space, it was necessary to lower the pressure in the combustion chamber to approximately 1 psi, which was the pressure that resulted when the starting flow choked at the chamber throat. The door at the end of the diffuser, which would blow open when the engine fired, closed off the test system for the start. The evacuation for starting was accomplished by the two-stage steam ejectors supplied by large steam accumulators. This system has operated essentially trouble-free on all RL10 test stands for the past 40 years, and it provides an essentially perfect simulation of starting and operating in space.

As the engine matured and consecutive firings became the rule, a way to close the door before the next start was needed. The stand was located approximately a quarter mile from the control room. To avoid someone having to make that hike to close the door, the crew rigged a bicycle chain and an air cylinder to close the door. Installed one night early in the program, it worked so well that it was added to all of the stands.

The RL10 flow schematic in Fig. 4.1 illustrates the elegant simplicity of the engine. There are few parts:

- A two-stage turbine drives the two-stage hydrogen pump.

- The single-stage oxygen pump is geared to the turbine shaft.

- Hydrogen flows from the pump to the chamber and exhaust nozzle jacket. In the jacket, the temperature of the hydrogen gas rises to −165°F.

- The thrust control, as a function of chamber pressure, bypasses the turbine, and the remaining hydrogen flows directly to the valve.

- Passing through the main hydrogen valve, the flow enters the injector.

- Oxygen from the pump passes through a valve to the injector.

The mixture ratio valve and the thrust control are adjusted on acceptance test.

In addition to the genius of John Chamberlain, which can be seen in almost everything the company did, Pratt & Whitney possessed another asset named Bob Abernethy. His invention that converted the J58 turbojet into a turbo-ramjet and opened the Mach 3+ flight regime is covered at the end of Chapter 3. Following an interlude as a Rhodes Scholar, where his field was statistics, Bob returned to FRDC to expand the department and to focus on the practical application of a statistical approach to the solution of many practical problems. One of the first efforts was the proper trimming of an RL10 for delivery. The objective was to set the controls with a minimum number of firings to ensure that the engine would meet specifications time after time in the field. Bob used a Monte Carlo simulation. As the name suggests, the method compares the random variation of how a complex system will react to each event to that of the spin of a roulette wheel or the toss of the dice. With a computer, thousands of firings of an RL10 are simulated, including all the potential random variations in things such as instrumentation error and propellant temperature. If the characteristic change in the engine to a change in the particular variable is included, the results predict the engine-to-engine and run-to-run variations. This complex modeling was impossible before the advent of the computer. A trim firing was performed, in which the chamber pressure and the mixture ratio were set within a bulls-eye that had a diameter equal to 1% of each of the parameter's specification goals. A check firing had to repeat within a slightly larger target. The Monte Carlo simulation predicted the probability of the engine being in trim in the field. The strategy proved so successful that NASA gave an incentive award to Pratt & Whitney.

A real advantage of the expander cycle is that the system is completely self-tending. It also produces the highest fuel efficiency of any engine ever developed.

The geared drive, which connects the hydrogen turbine to the liquid oxygen pump, deserves some mention here. This certainly was the first and may be the only such configuration. Most liquid oxygen turbopumps have a turbine that is independent of the fuel side. If the liquid oxygen pump has a serious cavitation, the pump overspeeds, sometimes to destruction. On the third flight of the McDonnell Douglas Phoenix single-stage to orbit test vehicle powered by four RL10-A5 engines, a malfunctioning valve in a vehicle ground helium system caused a simultaneous partial cavitation of the liquid oxygen side of all four engines at liftoff. The fuel side was stable. During liftoff, the vehicle began to stagger almost immediately due to the reduced thrust caused by the cavitation. Shortly after separation from the ground system, the cavitation stopped and the engines came up to full thrust, having cleaned out the propellant lines. The flight continued and concluded with a successful landing. The flight would have failed had it not been for the gears.

Carl Ring relates a story that unfolded several years later when he was running a demonstration of the RL10 for visiting dignitaries. A rash of failures of the speed pickup tangs had occurred; therefore, when eight seconds before scheduled shutdown, the speed dropped by 20%, Carl thought another tang failure was occurring. All other parameters remained unchanged, except for the gearbox pressure which dropped from 45 psi to 1 psi. Carl completed the run and had a normal shutdown. The visitors were impressed, and Carl proceeded to check the engine after they had departed. The side of one of the housings was missing a piece, and a 20% pie cut also was missing from the liquid oxygen gear. The ruggedness of the design allowed the engine to run for the final eight seconds with the gears skipping over the missing sector.

The thrust chamber is unique, and its development was a saga—compressed in time, but a saga nonetheless. Dan Sims was the assistant project engineer responsible for development of the thrust chamber. It was to be furnace brazed, with pure silver, and all of the tubes, manifolds, reinforcement bands, and a "Mae West" were assembled on a mandrel. The mandrel caused all of the parts to conform to the hourglass shape required of a rocket chamber and nozzle. The mandrel was loaded into a hydrogen retort, with the shaft protruding, and was rotated like a chicken roasting on a spit for the 20+ hours required for the braze cycle. The use of a static hydrogen retort technique for brazing had been standard practice for some time. The theory is that the active, hot hydrogen gas reduces any residual oxide contamination that might remain on the braze assembly. This was the first time a rotisserie function was added to prevent distortion and to promote uniform braze coverage. A measure of the high quality that the system achieved is that the centroids of ten sections throughout the length of the chamber and nozzle were within 0.015 inches. From the outset, the pre-braze assembly and all parts were handled with cleanroom techniques. Figure 4.3 shows the thrust chamber brazing retort.

This event is described in a few lines here, but for the first six months of the program, it required the round-the-clock attention of those involved. A description of the parts and their assembly may illustrate the difficulties and the momentous achievement of the people

*Figure 4.3 The thrust chamber brazing retort.
(Courtesy of Pratt & Whitney)*

involved. The function of the thrust chamber (Fig. 4.4) is to provide the space for the containment and combustion of the hydrogen and oxygen propellants delivered by the injector and the subsequent expansion of the hot gas. The shape of the bell is dictated by the requirement to provide efficient expansion.

*Figure 4.4 A view of the thrust chamber.
(Courtesy of Pratt & Whitney).*

The fire in the chamber is approximately 6000°F at 300 pounds psi. To provide the structure for containment required active cooling by the hydrogen flow, which later is expanded through the turbine to provide the power for pumping the propellants.

The double-tapered tubes were manufactured by LaFiell Company, which had developed the process for forming the tapered tubes for the production of golf club shafts. Mr. LaFiell took a personal interest in the rocket project. Likewise, Paul Burke, his manager, was an invaluable partner in its development. The thrust chamber expansion ratio of 40 was chosen as a balance between performance and weight. This was the highest expansion ratio ever attempted by anyone and led to the selection of the "pass and one-half" coolant tube concept for the design to maintain adequate coolant velocities throughout the length of the chamber. One hundred eighty tubes are located in the combustion section, each occupying a segment of 2° at each station in the combustion chamber up to the introduction of 180 interleaved half-length tubes. At that point, there are 360 tubes total, flattened and each to occupy 1°, which continues through the expanding nozzle. Figure 4.5 shows the details of the tubes and how they nest together.

Figure 4.5 The thrust chamber tubes and assembly. (Courtesy of Pratt & Whitney).

In the 1960s, our chief financial officer in Florida was a wag named Bob Foster. He was of modest height with a well-rounded physique, and he was devoted to golf. Among some, Bob had the reputation for being a tightwad. For example, he required business travelers to turn in any "green stamp" awards received on a company trip. One trick we had learned in the fabrication of the chamber was that bent paper clips and many elastic bands looped together could be used as an easy means to hold the 360 tubes onto the brazing mandrel before the bands and the turnaround manifold were fitted. After we were well along in the project, we took Bob on a tour of the chamber assembly room, which was closed off to keep the room clean. When he turned the corner into the room, Bob saw a partially completed chamber assembly with its 360 paper clips and more than a thousand elastic bands in place. Bob let out a whoop and said, " I wondered who was stealing all the paper clips and elastics out of the supply room."

During a hurricane, both Bob and I had taken refuge in the Camelot Motel in North Palm Beach. Our sailboat was in a slip located behind the motel, and I wanted to be near it to check the lines as the water rose. "Bob Oh," as he was known, wanted to remove his new car from Singer Island and parked the car in the inner court of the motel, in the shadow of the expected winds. He was on the second floor and was looking out the window at his new beauty as a piece of the roof blew off and landed on the car. It is terrible to see that emotion in one so "tight." My wife prohibited any further line tending of the sailboat after the water rose above the finger piers, but the boat survived without a scratch.

After the fitting of the tubes, manifolds, and bands on the mandrel, the complete assembly was installed in the braze retort, and a cover was welded on the end, with the central shaft protruding. The retort then was loaded into a standard gas-fired furnace, and the shaft was connected to the rotisserie, which slowly rotated the braze assembly throughout the brazing cycle. Pure silver braze is very strong but difficult to achieve. The temperature range between the liquid state (when silver melts) and the solid state (when silver freezes) is narrow and thus required careful monitoring of the thermocouples mounted on the rotating assembly, as well as almost perfect uniformity in temperature of all parts to be bonded. It also required scrupulous cleaning because an oily fingerprint is an abomination.

Typical practice at Pratt & Whitney was that after an engine was developed, responsibility shifted from the engineering to the manufacturing department in East Hartford. Thus, in 1962, the con was given to manufacturing. The FRDC Experimental Shop was "looked down on" by the shop experts in Connecticut, who would not listen to FRDC regarding the requirement for the ultimate cleanliness of the braze assembly. As a result, the East Hartford group never successfully brazed a chamber. Using the excuse that the low volume was not worth the trouble, production of the RL10 shifted back to FRDC.

Many attempts failed during the brazing development for the thrust chamber. One of the critical problems was to hold the bands in the proper position during the wide swing of temperature in the cycle. Dan Sims, who was responsible for development of the thrust chamber, worked on the problem day and night. He credits Ed Granville, manager of the East Hartford Experimental Shop (see Chapter 3), for the input that led to the final success.

Ed had suggested an external cage, with the proper thermal characteristics to hold the bands in place during the critical heat-up phase. I also remember Ed Lucas and his eventually successful struggle to clamp and braze the four parts "Mae West" around the combustion chamber.

The braze cycle took approximately 20 hours in the furnace, followed by another 6 hours to cool down and open the can. Therefore, plenty of nail-biting time occurred between each test. We would all stand there as one of the sheet metal mechanics ground off the seal weld on the end of the retort to see the results. The early failures were marked by distortions of the assembly, and if the failure was bad, a collection of tubes and parts sometimes lay in the bottom of the retort.

On the sixth try, the can was to be opened on third shift. Dan had left instructions that he should be called, so that he and Roy Clark, the metallurgist assigned to the job, could come to see the results. They arrived around 2:00 A.M. at the Florida plant, which is located several miles out the Beeline highway from the residential area near the coast. They went to the braze assembly room, where the mechanic was grinding off the closure. As the mechanic neared the end of the grinding, he partially pried the cover aside to allow everyone to peak inside. Dan looked inside and cursed, and Roy looked inside and immediately walked out of the shop. Loose tubes could be seen in the bottom of the can, signaling the latest failure.

After a reasonable period of sweating for Dan, the cover was completely removed, and inside was our first near-perfect chamber. It turned out that Don Cotrone, who ran the shop on third shift, had looked into the retort earlier and had seen the excellent results but decided to have some fun with Dan and Roy. The crew had put the extra tubes into the bottom of the retort, and they tacked the end cover back on and waited for Dan.

After a few expletives defining the ancestry of the jokers and the exclamations of joy at the great success after so many tries, Dan left to find Roy. He finally found Roy writing feverishly at his desk in the metallurgists' office. When Dan asked Roy what he was doing, Roy replied that he was getting ready to send out his resume. Roy was relatively new to Pratt & Whitney, and he explained that if he had had as many failures at his previous place of employment as had been experienced with the RL10 chamber, he would have lost his job. Dan explained that Pratt & Whitney did not work that way. Then Dan told Roy about the gag and what a beautiful chamber we had. Roy was greatly relieved.

The total freedom of not needing to look over your shoulder and having all of the necessary resources with which to work created a climate for everyone that yielded many solutions to many new and difficult problems in a short time. To start a totally new engine concept—with an untried fuel and in a unique new test facility that had never existed anywhere in the past—and to have the first engine running in nine months as long as the test stand tank capacity would allow were outstanding achievements.

The Voyager vehicle is one of the long list of accomplishments in space made possible by the RL10. In March 1998, I heard that the Voyager vehicle was approximately six billion miles from our solar system.

The thrust chamber tubes must be double tapered to follow the " Mae West" type of shape of the combustion chamber and nozzle. The smallest tube diameter occurs at the chamber throat. To achieve the intimate contact between tubes required for silver brazing, the local diameter of the tube, at any station, must fill the 2° sector allotted. The LaFiell Company produced the tubes by swaging Type 347 stainless steel tube blanks over double-tapered mandrels, with axial tension applied to the blank throughout the process. The Type 347 alloy was chosen for its high yield strength and good formability. It was readily available in high-quality tube blanks, ready for drawing.

The repeatability and accuracy of this process was a credit to the LaFiell Company and owed a great deal to the golf club shaft manufacturing experience. Pratt & Whitney received the straight, double-tapered tubes from LaFiell, with a wall thickness of twelve-thousandths of an inch. The first step was to bend the tube to the curve defined by the desired chamber shape. To provide control over the circumferential fit between tubes, the tubes then were filled with wax and "spanked" in book dies. The stamping provided the local flattening required to inter-leave the long tubes and the short tubes, as well as good dimensional control in the combustion chamber sections where the tubes essentially were round. Then the wax was removed, and each tube was flowed to classify its coolant flow capability. The important dimension to achieve the required tube-to-tube fit was the thickness of the tube in the circumferential direction of the complete assembly. These measurements were made at ten stations from the injector face to the throat for each individual tube, to an accuracy of 0.0001. With enough serialized tubes for three chambers, approximately 600 of each length, the computer selected the best possible tube-to-tube fit from the 6,000 dimensions, fitting the "pimples into the dimples," and separated them into chamber sets. A second constraint selected tubes in sets of five, and each group would have the same flow characteristics to ensure uniform cooling around each chamber. This system produced unmatched quality but was expensive. Tom Kmiec, RL10 engineering manager, says that the advanced process control techniques adopted by the LaFiell Company now produce tube uniformity that has eliminated the need for the costly statistical selection.

The ultimate, so-called 180-tube chamber has been covered previously in this discussion. The first few chambers were a prior design constructed with 144 tubes with a conical combustion chamber. The first firings to check the injector were made with lag-cooled copper chambers. The first cooled tubular chamber firing occurred in early May 1959. It was a disaster. The tubes in the nozzle skirt burned through in places, and a crash restudy of the heat transfer during the start transient was undertaken. Everything appeared to be correct, and no reason for the overheating was found. In the interim, to continue the testing, the areas of burn-through were insulated on the inside with furnace cement as a temporary solution to get through the transient. The study had indicated that at steady state, that particular section of the chamber was overcooled. Movies later showed water condensing from the products of combustion, flowing along the inside of the chamber. The hydrogen/oxygen flame is colorless and transparent.

These firings were done without turbopumps; propellants were supplied through test stand plumbing from pressurized tanks. The project group had specified hydraulically operated valves for propellant control in the stand because of their rapid response. The head of operations, which included facilities and test stands, had decided unilaterally that the hydraulically operated valves were too expensive. He had taken it upon himself to switch to much cheaper but slower air-operated valves. Don Hazard was an outside hire who had come from Wright Aeronautical Corporation, and he did not know how Pratt & Whitney operated. The valves were changed back to hydraulically operated ones. Later, in the same month, rated conditions were achieved, and all subsequent firings had no burn-through. The whole problem had resulted from the slowness of the air-operated valves.

The operation of the test stand was complicated and required the proper sequencing of various propellant, pressurization, and vent valves. Some of the stand valves began as air-operated valves. To ensure they operated on time, flags were attached to the stems, and a test engineer was assigned to check the timely operation with binoculars from a checkpoint located outdoors in the swamp behind the stand. The screened enclosure contained a seat and was about the size of a phone booth, and the structure inevitably was dubbed "the outhouse." Bill Creslein remembers "drawing the short straw" and being in the outhouse on a night firing early in the test program. Bill took a flashlight to check the corners of the enclosure for snakes before he was seated. He always wondered if the grunts and swamp noises he heard would be followed by the charge of a hog or alligator storming through the screen door of the enclosure. Bill tended to keep his feet off the ground.

The propellant injector was based on the concentric jet design that had been invented at the NACA Lewis Laboratory, before it became part of NASA. Abe Silverstein, who ran the lab, was interested in hydrogen, first for airplanes and later for rockets.

Figure 4.6 is a view of the injector and shows the oxygen spigot with a concentric annular hydrogen passage. In the NACA injector, the faceplate was sheet metal and depended on heat conduction through the face to the incoming hydrogen behind it for cooling. Welds to the posts near each injector element hold the faceplate in place. The major change from the NACA design was to replace the sheet metal faceplate with one made of a material called "Rigi-Mesh." It is a porous sheet, produced by hot rolling a series of screens together in a vendor's proprietary process. The porosity can be controlled by the process, and with the Type 347 wire, it produced a tough, almost "Damascus barrel" type of material. The original solid faceplates became distorted with the large number of firings now being performed. The use of the Rigi-Mesh material provided transpiration cooling, and distortion was eliminated. In the early days, Allen Tool Company was the only vendor that could make the injector weldment. An

Figure 4.6   The propellant injector.
(Courtesy of Pratt & Whitney)

employee of that company used a 2 × 4 to lever the Rigi-Mesh into place as he welded it. The problem was to maintain concentricity in the injector elements. Except for the switch to Rigi-Mesh, the injector has remained relatively unchanged over the past 40 years. Jim Ogan made one other important change by cutting the oxygen spigots, in the row near the tubular wall of the chamber, on the bias. The slight angle on each tip reduced the impingement of the oxygen on the thrust chamber cooling tubes and led to an improved life. A single engine eventually was operated at full thrust for 1,680 seconds, finally shutting down when the propellant tanks emptied.

Once the hydrogen-cooled bearing was proved, the pump program proceeded rapidly. The gas to drive the test turbine in B1 was hydrogen blowdown from a bank of tube trailers. To produce the required pressure rise at the lowest possible tip speed, the 304 and the early RL10 pump impellers had straight radial blades. As confidence was gained, the swept back design seen in Fig. 4.7 was adopted to improve stability. The hydrodynamic performance of the pump was achieved almost immediately.

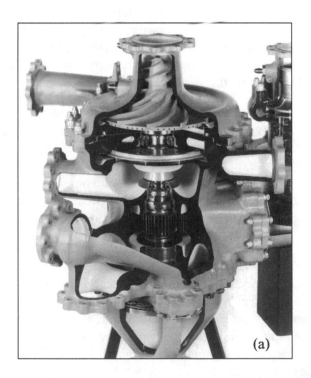
(a)

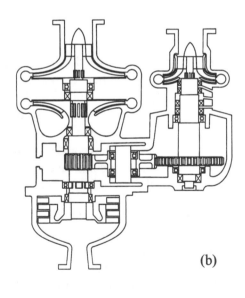
(b)

Figure 4.7 (a) Cutaway view of the turbopump. (b) Cross-section view of the turbopump. (Courtesy of Pratt & Whitney)

The liquid oxygen pump had a little four-inch single-stage stainless-steel shrouded impeller, which absorbed approximately 80 horsepower. It performed well from the outset. On the occasion of its first test, liquid nitrogen was substituted for liquid oxygen as a safety consideration. The density of liquid nitrogen is somewhat lower than that of liquid oxygen; however, if you encounter an unforeseen rub, at least there is no fire. To preclude having a large white exhaust cloud next to the test stand, the crew had extended the discharge pipe from the rig for approximately 100 feet outward to a pond behind the test stand. To make the job appear neat, the mechanic put a 45° bend downward, near the end, to follow the contour of the ground and to aim the pump discharge into the pond water. The stainless steel pipe was heavy, and the mechanic thought it would remain in place. On the first test, the pump worked well, and the blast of nitrogen out the end of the pipe was significant. Don Riccardi was watching the test on the outside and remembers that the stainless steel exhaust pipe rose up and wiggled similarly to a snake. The bend had caused enough reaction to make the pipe fly. The 45° elbow was removed, and the pipe was tied down for all subsequent tests. All of the liquid oxygen tests that followed were uneventful.

The first complete engine with turbopumps, the FX-121 (Fig. 4.8), ran in July 1959. The first test had a conservative duration of seven seconds, and every part appeared new after the test. Frank McAbee was the test engineer, responsible for the assembly and test of the engine. The engine ran so well that Frank soon was running it to the full duration of the propellant tanks. Many years later, Frank became president of the Florida division of Pratt & Whitney.

This engine was assembled with the 144-tube conical chamber, and it worked well. An alternative 180-tube chamber had been designed to provide additional turbopump power margin, and its first run was conducted in September. This first RL10 was hardly a thing of beauty. Dick Coar took one look at the ugly, braid-covered flexible joints in the plumbing, as shown in Fig. 4.8, and said the joints had to go. The decision was made to change to solid pipes, which were long and flexible enough to eliminate the joints if a special assembly procedure was employed. Lou Emerson's invention of a special seal and Hal Gibson's addition of Teflon® coating were the secrets of its success.

This seal uses a conical washer with a Teflon® coating, which is trapped in an angular cavity, half of which is machined into each of the opposing faces of the mating flanges. The angle of the washer in the free state is 30°. As the flanges are pulled downward, the washer toggles in the cavity, exerting intense pressure on the inner and outer edges. The calibrated pinch loads and the Teflon® coating provide excellent, gas-tight sealing in a lightweight joint. The Teflon® coating also protects the critical seal edges during handling and installation. It further slows seal shrinkage during cool-down to preclude transient leakage.

The assembly secret was to lightly assemble all of the pipes to their respective components and then pull down the flanges in small steps so that any residual misalignment, which experience showed to be minor, would be uniformly distributed among all of the joints. This is similar to the common procedure for tightening cylinder head studs. Our first problem was that delivered engines were reported as having leaks at NASA Marshall. It had been common practice there to take all delivered engines apart to conduct a detailed inspection. NASA was reusing the

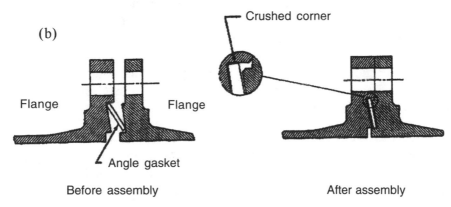

Figure 4.8   (a) The FX-121, the first RL10 engine.   (b) Angle gasket.
(Courtesy of Pratt & Whitney)

conical washers, and some of the Teflon® was lost after the first installation. We finally convinced NASA Marshall that exhaustive detail inspections had been done on all of the parts at Pratt & Whitney prior to delivery, and that their best guarantee for a leakproof engine was not to disassemble it after our acceptance test. They finally agreed, but it was a difficult sell.

The next difficulty occurred years later, when East Hartford was trying to assemble its first engine. Bill Creslein had gone to East Hartford to help, and in the afternoon he explained our rigid plumbing assembly technique. The foreman led Bill to the Pratt & Whitney assembly manual, which required that one end of the pipe be fixed and then the other end be pulled down. Thus, according to the foreman and the manual, they could not, under any circumstances, use our system. Bill gave them his telephone number at the hotel and said they could call him if he could be of any further assistance. Sometime later, Bill received a sheepish call to return. They had failed and were ready to try the FRDC procedure.

The seal on the oxygen pump shaft was designed with two carbon rub rings supported on bellows and two carbon split ring seals to separate the oxygen side from the hydrogen. The three compartments between them were exhausted to a vacuum, to prevent commingling of any minute seepage through the seals, to a pump on the ground, and overboard in space. We never experienced a fire from this seal, but suspicion existed that the carbons on the bellows were dancing around on the rub faces in the early engines. Don Riccardi, who worked for Bill Creslein, set up a seal and its mounting bellows, together with a simulated rub plate in a drill press. Sure enough, at the proper speed, the high-speed movies showed that the seal assembly would shimmy like the shimmy of a hula dancer. It was a case of stick-slip motion exaggerated by the softness of the bellows. Anti-rotation pins and later a spline corrected the problem.

The material selected for the first gears was Waspaloy®, which was known to be compatible with hydrogen. It was a tough nickel alloy originally developed for cast turbine blades and had been forged successfully for the first time in the 304 program. Gear rig testing passed with flying colors; however, the testing had been run in an air environment. When run in an engine, the gear box environment was gaseous hydrogen and after running for a while, the gear teeth tended to weld together. The hydrogen environment evidently was super-cleaning the teeth. A change to a conventional gear alloy, 6260, with a standard molysulfide painted coating did the trick. Subtle differences in material properties, resulting from the hydrogen and cryogenic conditions, were becoming increasingly important.

The rotational speed of the turbopump on the engine was measured by a series of pins on the hydrogen shaft that passed by a pickup coil. On the first attempted run, the speed reading in the control room indicated that the engine was accelerating at a much greater rate than had been anticipated. Furthermore, the engine tended to overspeed and was shut down almost immediately. We were puzzled because the numbers did not make sense. A mechanic went to the stand, turned over the engine by hand, and over the intercom called out each rotation. The counts agreed exactly with the correct number, and the mystery deepened. It turned out that the pins, at high speed, were each giving two pulses—one when arriving, and one when departing the coil position. Tom Pace from Instrumentation recognized the problem, and he added a clipper and a clamper to the pulse measuring system as the solution, and engine testing proceeded.

One of the first problems, which demanded a major effort at the outset, was the requirement to achieve pump cool-down within 20 seconds of the start signal. Being a second-stage engine, the start signal did not come until the first stage had separated. All of the propellants, which had to be dumped overboard, cost dearly in payload; therefore, it was important to minimize the quantity. Also, the vehicle was coasting and slowing down during this period; therefore, it was important to shorten the time.

In the initial design, an overboard dump valve was provided at the exit of the pump to cool the working parts and prevent unsteady flow during engine starting. A prestart period of 60 seconds was the best that could be achieved with this system but permitted engine testing to proceed. Sam Arline, APE responsible for systems, and his group finally found a solution to the rapid cool-down after six months, a hundred cold flow tests, and eight valve configurations. By changing to a sleeve valve that closed in two steps at the exit and adding a bleed valve to the interstage, the pump reached the necessary low temperature within the required 20 seconds. I can remember the elation felt by everyone when Sam called from the test stand to say they had achieved the time.

Another problem encountered on early tests occurred at shutdown. Occasionally after a successful run, the fuel pump inlet housing would blow off. High-speed instrumentation indicated that a pressure spike was occurring because the cool-down valves, which also functioned as vent valves, were not opening as rapidly as necessary. Pressure boost added to the valves resolved the problem.

The losses in payload, even with the shorter period, remained significant, and this eventually led to a system that precooled the pumps on the ground with liquid helium before launch to minimize the in-flight requirement. Unfortunately, with the secondary failure of a check valve, the cold pumps aspirated air. This air froze in the impeller and led to the only two in-flight failures the engine experienced in 40 years. The latest system employs redundant explosive valves to provide positive shutoff.

In early tests of the engine, the thrust control had a simple flapper valve in the servo system that produced an unstable feedback, causing thrust fluctuations. An inventive test engineer in Frank William's control group, Chuck "Stick Shift" Lagrone, devised a simple change that corrected the difficulty, and he accomplished it almost overnight. Chuck reasoned that the flapper was bouncing on the control orifice and that the pressure change it produced was nonlinear. Using most of the same parts, Chuck fashioned a control with a shear valve in place of the flapper, eliminating the instability. This is the only engine in which Pratt & Whitney supplies the control system, and there were lessons to be learned.

Anyone familiar with the early stages of rocket engine development knows that most test firings occur on the third shift. Probably after a day and an evening of struggling for perfection, reality sets in and the decision is made to go. Normally, two or three test engineers were present at the test firings to monitor the dozen or so strip charts that recorded the critical parameters. One night during the control stability testing, Hal Gibson got ready to run at 3:00 A.M. and was the only observer on hand. If Hal stood back far enough, he could see all of the critical charts—at least, he could see if the pen moved quickly, indicating a problem. Therefore, Hal gave the order to proceed. The engine stabilized at full thrust and operated smoothly for six minutes without any instability. After the test, Hal reviewed the data and realized that the control had failed at the start and had been stable because it was full open. The six-minute test had been run at 19,000 pounds of thrust, 4,000 pounds above the rated 15,000 pounds, without distress. Hal could see that the pen was steady, but he had been too far away to read the numbers. When his expected "reaming session" for running the test without adequate coverage was completed, Dick Anschutz called Hal into his office to thank him for the unintended prototype test of an up-rated RL10. Two engineers were present for all future tests, and red and green limit lines were added to the strip charts.

Looking back on that period, aside from the constant threat of cost overruns, things proceeded well. By November 1959, having made the first firing only four months earlier in July, a single engine was fired ten times for a total of 35 minutes. All of the technical problems were yielding to seemingly positive solutions.

The money problem resulted because the original estimate had been too low and because the total organization had been raised on fixed-price philosophy. This was Pratt & Whitney's first cost-plus contract. We believed that our job was to tackle any problems as soon as they surfaced, without consideration about whose responsibility it might be. We had not learned to wait and put changes into the contract. It was noticed that Convair, which was an old hand at the cost-plus business, tended to "keep its head down and point left" in times of trouble. I was summoned to Washington to meet with the deputy chief of the Centaur office, Bill Shubert. Bill was a former Navy commander recently assigned to NASA, and he "read me the riot act." He said ours was a ridiculous cost and that he had developed a rocket engine for the U.S. Navy for a total project cost of $2 million. Properly chastened, I returned to Florida, determined to re-examine the program and eliminate everything that was not absolutely necessary. I can remember the project meeting where we reviewed everything on order, and the ignition rig was mentioned. The vendor making the rig was having a manufacturing problem, and thus the rig was late. We already had more than 100 successful firings at that point; therefore, the consensus was that the rig should be cancelled. The one dissenting voice came from Leo Corrigan from design analytical in the back of the room. Leo said "I don't think you should cancel it. The engine is not lighting on the same spark every time." When the laughter from the assembled team subsided, still stinging from the lashes received, I said, "Who cares which spark it lights on, as long as it lights?" Then we proceeded. As we were to learn the following November, we should have listened to Leo.

In January 1960, *Aviation Week* carried a story that NASA would soon solicit the design and development of a 200,000-pound thrust, hydrogen/oxygen engine for the second stage of the Apollo. With the successful RL10 experience under our belts, we thought that Pratt & Whitney was in an excellent position to do this job. Dick Coar took a gang of us to East Hartford for three weeks of intense proposal writing. Having lived in Florida for three years, I vividly remember my return to the snow. The engine was named the RL200 (Fig. 4.9), and it employed a shunt expander cycle similar to the RL10 but with the addition of separate gear-driven inducer stages for both propellants to ease the vehicle propellant supply job. Because not all of the hydrogen being pumped by the first stage was needed to drive the turbine, only part of the flow was pumped in the second stage to provide for jacket pressure drop. The balance was shunted directly to the injector.

The RL200 installation sketch in Fig. 4.10 shows the low-speed geared inducer in front of the hydrogen pump, approximately in line with the thrust control. The NASA requirement was that the engine hydrogen pump must be able to handle the cryogenic fluid with a prepressurization of 130 feet of head above boiling, to be supplied by the vehicle. With the low-speed geared inducer, the engine requirement was reduced to 30 feet. It is a shame that this engine design was slated not to be developed. It is likely that a big RL10 would have had a major impact on the future of space.

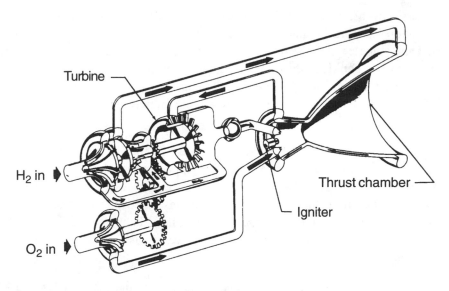

Figure 4.9   Schematic of the RL200 engine.   (Courtesy of Pratt & Whitney)

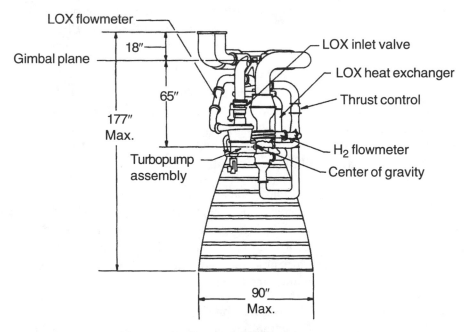

Figure 4.10   Schematic of the RL200 installation.
(Courtesy of Pratt & Whitney)

In mid-March, the proposal was delivered. We had high hopes because we were the only one of the three competitors that had significant liquid hydrogen experience, and we had a successful hydrogen engine under development, offering a superior design that met or exceeded all requirements. In due course, it was announced that the Rocketdyne division of North American Aviation Corporation had won the project, with an estimated development cost of $44 million. Our estimate was $138 million. Aerojet Corporation, the third competitor, had bid $66 million. I wish we could see the true development cost after the fact. I would bet that the J2, as it was called, cost significantly more than $44 million. Aerojet cried foul. We returned to the RL10 development. We had a lot to learn about competing with the chosen instrument of NASA.

An ironic twist to this story occurred several years later. At the Goddard Award dinner, I was dancing with my sister when Wernher Von Braun tapped me on the shoulder. He radiated charm as always, and my sister Rae was amazed that I was on a first-name basis with the celebrity. Wernher said in his thick accent, "Deek, it was our trip to Florida to see the RL10 firing that gave me confidence to use hydrogeen for Apollo." I can never think quickly enough on my feet to say the right thing at the moment. However, I wished I had said that I was sorry we hadn't been able to help him more.

By April 1960, we were ready to try an in-house Preliminary Flight Rating Test (PFRT). The PFRT is a holdover from aircraft days and signified when an engine was ready to begin flight test. Build 9 of the FX-121 completed the required 20 cycles, accumulating 2,442 seconds of operation on April 30. It was a joyous occasion. Unfortunately, the official PFRT did not occur until November 1961. The U.S. Air Force project office and all personnel were assigned to NASA under Dr. Hueter on June 1, 1960. This was the first step in the NASA takeover of the program.

On August 10, 1960, NASA contracted with Pratt & Whitney for an up-rated version of the engine, the LR119, to develop 17,500 pounds of thrust and to be used in a cluster of four on the S-IV stage of the Saturn. The stage later was changed to the S-IVB, which used six of the Centaur engines. This configuration was to be launched approximately a half-dozen times but was not slated for any follow-on application.

We made a major mistake in negotiations with NASA for an up-rated RL10, which often occurs among those too close to a recent development and who know all of the things that might go wrong. "Del" Tischler of NASA, who has always been our friend, tried to convince us that a modest program should be able to up-rate to 20,000 pounds. Our proposal was conservative and based on potential difficulties. As it turned out, little more than turning up the wick was needed to reach 22,000 pounds and higher.

On August 16, the first XLR-115-P-1G was delivered to Patrick Air Force Base. The "1G" in the name stands for "Ground Test," and the formal transmitting document, the DD250, is a comment on the times. It was a one-page document with two signatures on it: my signature, which certified that the required acceptance tests had been performed, and the signature of J.C. Jennings, the government representative, which certified that he and the government accepted the engine. Our old friends, Brandon Transfer & Storage Company, did the delivery. They had transported the 304 hydrogen expander engines in 1957 when testing in Florida first began (see Chapter 3).

The first meeting of the NASA Propulsion Assessment Team was held at FRDC on October 25, 1960. This marked the beginning of what Von Braun was to call "...the penetration of the contractor."

In November, my family accompanied me on an open house at the plant and saw an RL10 there in the flesh. Figure 4.11(a) is a photograph of the group peering into the thrust chamber and evidently is a good human-interest shot. Pratt & Whitney has used it many times, most recently in its fortieth anniversary brochure. Figure 4.11(b) is a recent photograph of a little boy named Jonathan, age 2, reaching toward the RL10 on display at the Children's Museum in Durham, North Carolina. Jonathan is the son of the little blond girl holding her parents' hands in Fig. 4.11(a).

Construction was nearing completion on a new test stand, E5, which was an attempt to simulate the Centaur launch vehicle. Prior to this time, all RL10 engine firings had been conducted on horizontal test stands. The E5 was our first vertical test stand. It had two engines mounted side by side in the vertical position and was to permit the first test with vehicle plumbing and the Centaur tank-mounted propellant boost pumps. Al Gardner was visiting, and he and I witnessed the first simultaneous firing of two RL10s on November 6. The test stand timer malfunctioned, and the firing was shorter than planned; otherwise, everything seemed fine. The next day, the test was attempted again. This time, one engine fired, but the other did not. The hot engine lit the unburned propellants from the other, and both were severely damaged. The cold one was reduced to small bits, and the firing engine was damaged beyond repair. I had not witnessed this test, and I could hardly believe the phone call from the test stand personnel because I had seen the good performance on the preceding day.

In those days, our test stands were much less automated than they eventually became. The test conductor was required to go through a lengthy countdown to put all of the necessary valves and switches in the correct position. The data check showed that the pressures at the pump inlets were far below specification. On the cockpit voice recorder, the operator cursed when he realized he had neglected to turn on the tank-mounted propellant boost pumps. These pumps had been on the preceding day, and the engine would not be expected to fire without the pumps operating. It seemed as if the problem had been found. Fortunately, relatively little damage had occurred to the test stand, and it was repaired quickly and the next test was planned for January. I can remember the trip to Washington to present our findings. It was a mixed audience of the Air Force, which was leaving the project, and NASA, which was taking over the project. There was no disagreement, and we were sent home to try again in January.

On December 20, I was assigned to be the chief of advanced technology for FRDC, and Dick Anschutz took over as project engineer for the RL10.

In January 1961, before a large audience of Air Force, NASA, and Pratt & Whitney people, the second explosion occurred; however, this time, there were no extenuating circumstances. The voice recorder heard the "Holy cow!" from the test conductor at the time of the explosion. A short time after the second E-5 explosion, an injector blew up on a horizontal test stand. The steam ejectors had not run long enough on the prestart, and evidently the pressure

(a)

(b)

Figure 4.11 (a) Open house at the FRDG plant, with my family peering into the RL10. (b) Years later, my grandson Jonathan, the son of the little blond girl in Fig. 4.11(a), reaches toward the RL10 at the Children's Museum, in Durham, North Carolina.

was not low enough to prevent flashback through the injector face. However, we did not yet know that such a thing could happen.

Dick Coar, who was now chief engineer of FRDC, devoted almost all of his time to solving the problem, and he organized the group into teams that focused on specific parts. I was pulled back into the project again and assigned to correct the ignition problem. Now it was time to build that ignition rig I had canceled in our economy drive almost two years before this time. John Chamberlain was on my team, as well as Frank Williams and Hank Vaughn.

I have a distinct memory of being in the control room in E-area on a Saturday afternoon when we went through a start sequence with a slight hydrogen gas purge on the fuel side of the injector to preclude any possible backflow of oxygen. The thing was fireproof and simply would not light, although many attempts were made.

By this time, approximately 230 successful firings had been made in the horizontal test stands. As it turns out, all had been flukes. This engine starts with a liquid oxygen lead. In those 230 firings, a small quantity of oxygen had flowed back into the injector fuel elements before the hydrogen flow started. When the hydrogen gas came into the injector manifold, it mixed with the minute quantity of backflowed oxygen, and enough of the premix issued through the elements near the spark to give 230 successful lights. Another factor, which hid the backflow, was that until that time, no internal injector explosions had occurred on the horizontal stands because the 1 psi pressure at start was too low for the flame to propagate back through the injector elements. They were small enough to act like the screen on a miner's lamp.

At the time, the theory was that tipping the chamber from the horizontal to the vertical position was reducing the tendency for the oxygen to flow back into the hydrogen side. Last year, John Chamberlain told me he had come to believe that the most significant cause had been the shift to the new test stand, with new steam ejectors that were run for a longer duration than those in the horizontal test stands.

John's genius showed again when it was time to find a reliable correction. John reasoned that slightly downstream of the liquid oxygen valve would be a reliable source of gaseous oxygen after the valve had opened, continuing until the time that the fuel valve opened and the ignition signal was given. John combined this oxygen source with part of the starting hydrogen flow, in a central pilot, concentric with the spark. As a little side invention, to measure the mixture ratio in the pilot to ensure that it was in the proper range, John installed a small whistle on the igniter tube and recorded the pitch. A higher pitch meant more hydrogen and less oxygen, and lower meant vice versa. Having John function as a consultant to all projects rather than being assigned to one was an excellent call by the engineering management, spreading John's remarkable talents across the company rather than on only one project.

This ignition system has worked successfully, essentially unchanged, for a thousand firings in space during about 40 years. Most recently, a simplification in the system has removed the gaseous oxygen valve and induced the recirculation of oxygen back toward the spark by modifying the hydrogen jets directly around the spark. This system also has been functioning well and now lights on the first spark every time.

The other major change came from Dick Anschutz's team. It was the modification of the control to a ladder system, which required that each successive event had to be confirmed before the next step could be taken. Prior to that time, it had been strictly a scheduled control where a timer ticked off each event that proceeded, without regard to whether the previous step had been accomplished. This simple system mirrored the system used by the Centaur vehicle at the time. Dick Coar, who in his earlier days had worked in the control group and recently had been busy with the current jet engine project, the J58, was appalled when he realized that the simple scheduled control was used in the RL10. He also caused the test stand system to be modified to provide an automatic advance to a safe shutdown in the event of a minor glitch, as well as a rapid abort for a major problem.

Development work proceeded on one of the major remaining system problems. During the start transient, thrust would overshoot to 18,000 pounds, 3,000 pounds over the rated 15,000 pounds. Excess turbine power when the engine starts at a high ambient temperature was beyond the limiting capacity of the simple control. This was acceptable with the engine because it was a single spike and well within the structural capability. However, it evidently was not so for the vehicle thrust structure, and we proceeded with work to reduce the overshoot. The addition of a small volume and an orifice provided anticipation in the control circuit, which limited overshoot to an acceptable level for the vehicle. This was another example of our lingering fixed-price philosophy.

NASA continued the "penetration of the contractor," and Leonard C. Bostwick from Marshall and his group of assistants arrived at FRDC to monitor the program. Their direct approach of questioning the project people and supporting staff was disruptive to the program. In June 1961, W.L. "Bill" Gorton, our old boss, was named general manager of FRDC. He corrected a major initial organizational mistake by bringing with him key support staff from East Hartford. He also brought Bruce Torell, who most recently had salvaged a troubled program at Pratt & Whitney in Canada and who became my boss as program manager for the RL10. Soon after Bruce arrived, he and Bill flew to Huntsville to confront Wernher Von Braun to explain that it would be impossible to complete the development with such interference. Wernher recognized the strength of this team and their argument, and he agreed to "call off the dogs." In return, Bill and Bruce resolved to meet all requirements and to institute a weekly meeting to keep Bostwick and company up to date. Internally, it was dubbed the "happy hour." The interference decreased, and work proceeded rapidly. Bruce did a yeoman's job of defending the project from the onslaught of the NASA hordes. He later became president of Pratt & Whitney.

Working with NASA was a learning experience for Pratt & Whitney. NASA introduced us to significant improvements, particularly in instrumentation and automated control of the test stands. Over the years at Pratt & Whitney, engines had had a lower metabolic rate than that of rockets. With an experienced crew, the time-honored technique of a checklist and a clipboard to read gauges and a manual countdown was the system at Pratt & Whitney. NASA's rocket engine experience had taught them that a much more automated test stand control system was required. The number of performance engineers also was doubled to handle the large increase in data from each test. The gentle nature of the RL10, which operates similarly to a jet engine, was the only reason that Pratt & Whitney was able to use the old system as long as it did.

NASA also instituted requirements for detailed planning, which conflicted with the "hands-on" system we had learned as project engineers. As our small group was breaking new ground, we would plan today's next step at a short morning meeting, often held on the test stand, based on the test results of the preceding night. It was impossible to publish a detailed plan and continue to move as quickly as we did. The need for a detailed plan increases as a program matures and as more people become involved, but not at the start. The rapid growth of the RL10 technology, which was totally new in so many dimensions, could not have happened with the formality that NASA sought. The requirement for sealed pre-run declarations of the expected results prior to every engine test was a bit much. However, the engine was a natural, and the official PFRT of the RL10-A1 was completed on November 4, 1961, in only six days of testing. Bill Creslein was the test engineer, and Frank Williams was the project engineer.

In the fall of 1961, Gordon Titcomb, who later became executive vice president of the commercial division of Pratt & Whitney, joined the project, taking over my job and enabling me to return to the advanced technology assignment, which will be covered in Chapter 5. Gordon tells a story about the day that Frank Williams called from the test stand to say that they were ready to begin the PFRT for the A1 but that the NASA representative, who was supposed to witness the test, had not arrived. Gordon remembers with relish his call to Bostwick. He said the test was about to begin and would proceed even if the NASA representative were too busy to attend. Frank remembers the squeal of brakes and flying stones as the man arrived just before the countdown reached start.

I am particularly fond of one memory, which falls into the category of "Do as I say, not as I do" and took some time to play out. A fundamental difference in philosophy concerned the use of component testing before a complete engine test. NASA wanted every part tested, as a component, to exhaustion before committing to an engine test. The Pratt & Whitney philosophy was that complete engine testing was desirable as soon as possible because many hidden interactions occur among engine components. These interactions often are hidden and unexpected, not simulated by component tests. If allowed to continue too long, the interactions may hide a real problem. A neat closure occurred in the early phases of the Space Shuttle development. One of the remaining major component test stands at Rocketdyne was for testing the liquid oxygen pump. After the test stand blew up in an early test, the press release from Marshall indicated that the test stand would not be rebuilt because the complete engine was a better simulation of real-world operation. NASA had learned something as well.

The second version of the engine, the RL10-A3, completed its PFRT in 1962, again in less than one week of testing. The major differences in the A3 from the A1 were in the pumps, which had improved suction performance to enable the Saturn S-IVB to operate without boost pumps, as did the Centaur some time later. The second change was to a smaller throat at a higher chamber pressure, yielding a higher nozzle expansion ratio within the same length and a six-second improvement in specific impulse from 438 to 444. The original specification was 420 seconds.

In August 1962, the capability of operating the RL10 over a ten-to-one thrust range was demonstrated. An excellent film shows Hal Gibson running such a demonstration in one of the new single-engine vertical test stands. A movie shot taken from the rear of the engine shows water and frost forming on the inside of the nozzle. The flame is absolutely transparent. In February 1963, the ability to operate at a very low thrust without pump rotation, strictly on tank pressure, also was demonstrated. The RL10 seemed to be able to do almost anything.

Bruce Torell became convinced that an additional liquid hydrogen storage tank was needed to smooth the bumps in the liquid hydrogen supply from the "Papa Bear" plant and the RL10 test consumption. He was unsuccessful in selling the idea to Wernher Von Braun, so he had Bob Abernethy build a super-Monte Carlo model of the complete liquid manufacturing, transportation, and consumption in test, as well as the estimates made by people all along the way. The simulation was so successful that NASA quickly changed its tune and supplied the new tank, as well as the additional tractors to move the Dewars that the study had shown to be a bottleneck. RAND Corporation, the U.S. Air Force "think tank," said it was the largest Monte Carlo it had seen, and the Air Force adopted the system for all hydrogen logistics and consumption in the country.

In 1980, I was invited to be a consultant to the Aeronautics and Space Engineering Board of the National Research Council during its review of the NASA plans for liquid rocket research and technology for the next decade.

This was a time when the Space Shuttle Main Engine (SSME) was in the depths of development struggles. The plan presented supposedly would lead to a new, high chamber pressure, hydrogen/oxygen engine in the RL10 thrust class. This was the next serious challenge to the RL10 of which I was aware, and it was poorly timed. As discussed in Chapter 5, no real advantage is offered by high chamber pressure in upper-stage engines; rather, there are plenty of difficulties. The board, which included old NASA hands Bob Gilruth and Abe Silverstein, found instead that the RL10, which by that time had experienced more than 10,000 firings and one and one-half million seconds of operation with a perfect flight record, should be modified and improved to do all of the missions in its thrust class. The RL10 had been missed by another bullet.

By September 1963, approximately 3,300 firings of the RL10 had been completed on test stands at FRDC and approximately 550 firings in space, counting two engines for each flight and multiple firings on some missions.

When I needed to find for this book a picture of the RL10-A3, an engine built over many decades both for NASA and as a commercial product, I could not find one in the Pratt & Whitney files. Bill Creslein suggested I contact Bob Foust who might have a high-quality photograph. Bob had been involved with the birth of the RL10 in 1958 as a test engineer and later was manager of the engineering program for the engine during its mid-life, retiring in 1993. Bob could write a separate book about the struggles to keep the engine alive. Bob provided the photograph for Fig. 4.12, an up-rated version of the engine introduced in 1986, the RL10A-3-3A. In this model, the expansion ratio of the exhaust nozzle had been increased from 40:1 to 61:1, thereby raising the specific impulse to 444.4 seconds from 420. The thrust also was increased by 10% to 16,500 pounds.

Pictures and descriptions of the various engine models and their missions can be found on the Internet at www.prattwhitney.com.

The impact of the RL10 engine on the accomplishments in space is not immediately apparent. Since its first flight in November 1963 through 1999, 76 communication satellites have been placed in synchronous orbit that permit worldwide voice and data transmission, including real-time television. More than 20 different types of satellites from a variety of different countries have been launched over this period and serve as the backbone for intercontinental connection of the world. The RL10 has played a part in all of the following:

Fig. 4.12 The RL10A-3-3A engine. (Courtesy of Pratt & Whitney)

- Prior to the Apollo mission to the moon, seven lunar-lander vehicles accomplish the precursor Surveyor function to select the landing spots

- Ten planetary fly-by/orbiters: Mercury, Venus, Mars, Jupiter, Saturn, Uranus, Neptune (Mariner, Pioneer, Voyager, Cassini)

- Two Mars landers (Viking)

- Two solar probes (Helios)

- Five astronomical observatories (OAO, HEAO), including the Hubbell telescope

- One technology demonstrator (ATS)

- One Venus atmospheric probe (Pioneer Venus 2)

- Four scientific experiments (CRRES, SOHO, SAX, EOS)

- Three weather satellites (GOES)

- Several classified Department of Defense payloads on the Titan IV

Figure 4.13 shows the latest model of the RL10, the B-2, on the J4 test stand at the Air Force test facility in Tullhoma, Tennessee. The two-position nozzle is fully extended. The carbon/carbon skirt that effects a 285:1 expansion ratio yields a specific impulse of 466.5 seconds. This engine, which is installed on the Boeing Delta III launch vehicle, made its first successful flight on August 23, 2000. The current backlog is 18 launches.

Figure 4.13   The RL10B-2 on the J4 test stand.   (Courtesy of Pratt & Whitney)

CHAPTER 5

High-Pressure Rockets—A Decade and One-Half Billion Dollars

New air-breathing engines at Pratt & Whitney evolved from the efforts of well-established Technology and Research (T&R) component groups that worked continuously to improve the performance of their particular components. Fans, compressors, combustion systems, turbines, and all other elements needed for turbofan engines were covered. Such groups existed both north (Connecticut) and south (Florida) to serve the respective commercial and government gas turbine engine requirements.

This system worked well in our established business, but no support for rockets was available. Having "gotten our feet wet" with the RL10, which borrowed heavily from the 304 engine (see Chapter 3), an adequate new technology base was needed for the emerging high-pressure staged combustion cycle that offered great potential for the next generation of rocket engines.

The original invention of the staged combustion cycle is credited to—again, who else—John Chamberlain. While studying ways to increase the chamber pressure level attainable with the expander engine cycle, John reasoned that it would be necessary to raise the turbine inlet temperature by burning some oxygen in the hydrogen working fluid before it entered the turbine. *Voilà!* The staged combustion cycle was born. Its similarity to that of an afterburning turbojet was not recognized until later. Bill Sens and Bob Atherton made significant contributions to the final cycle as the engine was developed.

All other things being equal, the expansion ratio of the exhaust nozzle determines the performance of a rocket engine. An increase in chamber pressure allows a higher expansion ratio to be installed within the diameter constraint of the stage. It also provides improved performance in booster stages where the expansion ratio usually is not limited by diameter but by the potential for overexpansion. The staged combustion cycle makes possible high chamber pressure without incurring the losses associated with the gas generator cycle, which is a holdover from Peenemunde. These losses are magnified as turbopump power requirements increase and preclude using the gas generator for high pressure.

The December 1960 tests of a 500-pound thrust motor began the technology program for the high chamber pressure rocket engine. This effort extended over the next decade and covered all of the components of an engine in three different thrust sizes. Investment in this work at Pratt & Whitney was almost one-half billion of today's dollars. The costs were shared by Pratt & Whitney, the U.S. Air Force, and NASA.

These early firings were of short duration. The chambers were thick, solid copper and depended on the thermal capacity of the metal, sometimes referred to as lag cooling, to absorb the heat generated during the brief firing. There was no active cooling. A bright green flash signaled that the test had been run too long. With good high-speed measuring and recording equipment, useful data could be obtained in a short test. These early tests showed that performance and stability were good but that injector durability was a major problem. The size of the test chambers was increased to 5,000 pounds of thrust and later to 10,000 pounds of thrust over the next year and a half. In approximately 40 firings and many alternative designs, good injector durability was demonstrated. Film cooling also was tested; it proved adequate but costly in performance.

A characteristic of operating at high chamber pressure is the significant increase in the metabolic rate of everything in the engine. For example, the maximum heat flux, which occurs at the throat of a hydrogen/oxygen rocket, grows from 10 Btu per second per square inch in the RL10 (which operates at 300 psi) to approximately 100 in a 3000 psi chamber. The unit is a measure of how much heat passes through a given area in a unit time, and a high rate can result in significant thermal stress. An equivalent rate in the F100 gas turbine first-stage turbine vane operating on jet fuel is 1.5 in its hydrocarbon/air environment. The heat flux is determined by the heat transfer medium and its temperature, and the oxygen/hydrogen system at 6000°F with excess hydrogen is extremely severe. If the surface temperature is established, the heating rate does not depend on how the surface is cooled.

The concept of repeated use for rocket engines drove the search for a long-life cooling system. Although convective cooling was acceptable in the tubular-wall stainless steel RL10 chamber, an order of magnitude increase in heat flux inevitably would reduce life expectancy. Transpiration cooling, in which a flow of coolant is bled through a porous surface against the heat flow and carries away the heat before it reaches the metal, had been applied successfully in other fields. What was needed was a practical way to apply efficient transpiration cooling to a rocket engine chamber. We are approximately halfway through this book, and the genius of John Chamberlain may begin to sound like "a broken record." However, John Chamberlain invented the radial wafer system shown in Fig. 5.1.

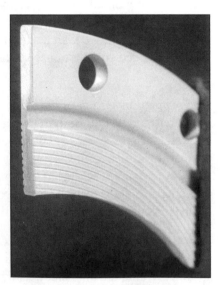

Figure 5.1 The 10K radial wafer system, invented by John Chamberlain. (Courtesy of Pratt & Whitney)

All of John's concepts have an elegant simplicity, and the transpiration wafer is no exception. In a flat copper wafer, small involute coolant passages are etched into both sides, in an opposing sense so that they crisscross on facing wafers. They spiral inward from the circular passage cut into the wafer, which acts as a distribution manifold. The fine, closely spaced grooves provide an effective heat exchanger that brings the injected coolant close to the melting temperature of the wall, thereby minimizing the coolant requirement. The chamber life becomes essentially limitless. The holes in the outer part of the wafer were for tie bolts, which held the test assembly together. Figures 5.2 through 5.4 show the chamber wafer assembly. Fully assembled and ready for testing, the 10K chamber was mounted on the E1 test stand in the north complex in mid-1963. The heavy hardware is a comment on the world of battleship test articles for high chamber pressure.

In July 1963, the 10K chamber produced 11,000 pounds of thrust and demonstrated efficient transpiration cooling at a chamber pressure above

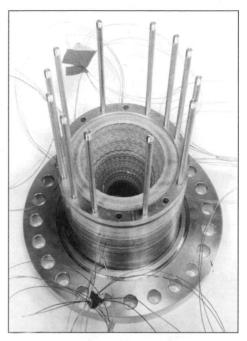

Figure 5.2 The 10K wafer chamber assembly. (Courtesy of Pratt & Whitney)

Figure 5.3 The 10K chamber. (Courtesy of Pratt & Whitney)

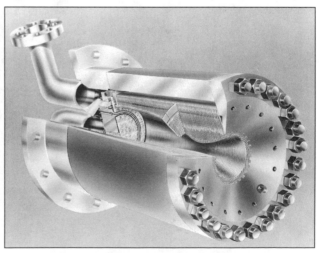

Figure 5.4 Cutaway of the 10K chamber, showing the relative position of the wafer elements. (Courtesy of Pratt & Whitney)

3300 psia. The transparency of the flame in the daylight (Fig. 5.5) hides the beauty that can be seen at night. This final firing of the series ran for 6.8 seconds, limited only by the size of the propellant tanks.

While this critical element was being developed and demonstrated, the other "tall poles in the tent" also were being addressed. Starting in 1962 and extending into mid-1964, first the fuel and then the oxidizer turbopump programs in a 50K size were underway. As with the cooling system, the high-pressure turbopump work opened the door to many new and difficult problems. Unfortunately, a photographic record of this work has not been found. However, there is a memory of Carl Comoli's struggle to make the turbine on the fuel pump hold together at high speed. Carl was the project engineer for the 50K turbopump program. Lou Dagne and Jim Sandy in the design group worked on these pumps and all other cryogenic pumps in all of the projects. The wrap of fiberglass strands over the tips of the turbine blades could not be a long-time solution, but it was satisfactory for completing the test series and demonstrating a pump pressure rise of 4390 psi. A total of 50 tests occurred in this part of the program, with almost 1,400 seconds of operation, and an effective axial thrust balance of the rotor system was demonstrated.

Figure 5.5 The 10K transpiration-cooled chamber, fired in daylight.
(Courtesy of Pratt & Whitney)

The oxidizer turbopump was a challenge. However, thrust balance and a pressure rise of 4770 psi were demonstrated in 32 tests, for a total duration exceeding 1,600 seconds. Later in 1964, the pumps were combined with a 50K transpiration-cooled chamber for pump-fed firings, with some durations as long as 17 seconds. Despite these great successes, the depth and number of troubles encountered foretold of a long and trying path. These difficulties were viewed as a mixed blessing because it would be great when we solved them and difficult for competitors to follow.

Also in the spring of 1964, we tackled the preburner, another key element of the staged combustion cycle. Staged combustion with its preburner and main chamber is similar to the main burner and afterburner in a fighter engine turbojet. The function of the preburner, where the pressure is approximately 5500 psia, is to generate the steam and hot hydrogen gas that drives the turbines. The distance between its injector face and the turbines is a matter of a few inches; therefore, a uniform temperature distribution is one of its main requirements. From late 1964 until the spring of 1965, the 10K preburner was tested by itself with many variations in design. Then it was coupled with a main injector and a transpiration-cooled chamber, and the first staged combustion tests were performed. After the proof of concept and configuration were demonstrated in the 10K size, hardware was fabricated in the 50K size. Staged combustion with a fixed area preburner, radial slot main burner, and transpiration-cooled chamber was demonstrated in September 1965. The work described to this point included 296 tests of various major components in battleship-type hardware up to 50K size for proof of concept, and it was supported by Pratt & Whitney and the U.S. Air Force Rocket Laboratory in Edwards, California. Captain Dick Weiss and civilian Technical Manager Jerry Sayles were enthusiastic supporters of this work. Although the military personnel involved in the project changed assignments from time to time, Jerry followed the high-pressure engine work at Pratt & Whitney from start to finish, and he once remarked that all advanced rocket work was not necessarily done in California. Jerry shared my addiction to sailing and, in his retirement, enjoyed much of it in the Pacific.

Captain Ernie Braunschweig took over the management of the project for the Air Force. With his superior technical understanding and his firm approach, he established an excellent working relationship. Ernie explained the "Golden Rule" to me during a contract negotiation: "He who has the gold makes the rules." For this contract, we were granted a one-dollar fee, and in passing through the Los Angeles airport on the return trip home, I lost a dollar in a change machine. Many jeers awaited me in Florida concerning my business acumen.

While this experimental program was proceeding, an analytical study of the benefits of the high-pressure engine to the vehicle contractors and to the Air Force and NASA was a major part of our effort to promote the concept. This work included more than engine performance studies. For the first time at Pratt & Whitney, detailed vehicle operational analysis was tailored to the specific requirements of the particular vehicle contractor as those requirements evolved with time. Joe Sabitella, first in East Hartford and later in Florida, was the pioneer in the company to approach rocket engine installations from the viewpoint of the vehicle. We had discovered that the way to attract the attention of the vehicle contractor was to present our data from its viewpoint.

Bob Atherton was a "Renaissance man" with a broad scope and many interests. On a previous assignment, he had worked on an engine that used liquid metal as a working fluid. The program was called NEPA, which stood for Nuclear Engine for the Propulsion of Aircraft. Therefore, staged combustion and liquid hydrogen seemed straightforward to Bob. In addition to his great hands-on experience, Bob also had a feeling for advanced system analysis that was to stand us in good stead. Bob already had made significant improvements to the staged combustion cycle.

From the outset, Bob and I worked well together, similarly to the way Dick Coar and I had related to each other over the years. Bob's strong analytical capabilities, combined with his hands-on experience in the field of nuclear propulsion, made him an ideal partner in the attempt to secure a piece of this business for Pratt & Whitney. Bob was more than extremely skilled—he was also a nice guy.

Previous competitions had shown that we were not on a level playing field, and we concluded that it was essential to try to pull the definition for the future over to Pratt & Whitney's area of expertise, which was long-life reusable engines. To give a face to the proposed product, a mockup was constructed to show how such an engine might appear. It was named the RL20 and is shown in Fig. 5.6.

Figure 5.6 Mockup of the RL20P-3 high-pressure engine. (Courtesy of Pratt & Whitney)

The engine was designed to produce 250,000 pounds of thrust at sea level, with an expansion ratio of 20.5 and an exit diameter of 3 feet. The mockup was completed in time for display at the United Aircraft Corporation Exposition in 1963, which was held at the Los Angeles Coliseum. All United Aircraft products were on display at the exposition, and it was the first public presentation of the high-pressure engine philosophy. The mockup, which had been lovingly crafted by our Model Shop, was handsome and was the center of attention in that hotbed of the space industry. This exposure was our opening salvo to identify the high-pressure rocket engine with Pratt & Whitney.

Rocket display mockup construction was the favorite job of the crew in the Model Shop, but the crew usually worked on more mundane engineering mockups employed in the fitting and external dressing of the pipes and wires. The mockup was constructed carefully of solid cherry wood, and the finish was something to behold. Unfortunately, the mockup had been dropped during unloading at the airport in Los Angeles. When the box was opened at the airport, the mockup had been broken into many pieces. The exhibition was scheduled to open in less than a week, and the "old boy" network that exists in the aircraft industry sprang into action. At that time, the Saberliner Division of North American Aviation was located next to the airport. Pratt & Whitney built the engine for that aircraft, and the two companies were on good terms. Somebody at Pratt & Whitney knew somebody at North American Aviation, and our disaster was taken into the North American Aviation Model Shop and beautifully restored in only three days. We were forever grateful for the monumental effort. Likewise, the Rocketdyne Division of North American evidently was impressed by the RL20, and I was offered a job by Rocketdyne at the show. In my refusal, I commented that I could not leave the work at Pratt & Whitney because it was too engaging. It was the truth.

The RL20 mockup was shipped next to the annual Air Force Association meeting, where it was viewed with interest by the Washington space world. This first mockup had successfully introduced Pratt & Whitney's cutting edge technology efforts to the rocket world.

To establish the concept of long-life reusable rocket engines, the brochure "Space Transport Engines" was published by Pratt & Whitney on September 20, 1963, and is included here as Appendix B. As with the mockup, the intent was to recast the rocket field into a place in which Pratt & Whitney could compete successfully. The brochure contrasted the then-current engines with the high-pressure staged combustion engine. The brochure was a great success and won an award as the best commercial house organ produced by industry for the year. Six years later, we learned that the concept expressed in the brochure had a significant impact at the upper levels of NASA.

The second paragraph of the brochure states:

> To provide an economic space transportation system, it appears that the next generation of launch vehicles will be developed as reusable systems. Engines for these vehicles will be significantly different from the throw-away types in current use, and must possess many of the characteristics of the engines in use in today's jet aircraft. Stability over a wide range of operating conditions, variable thrust to permit vehicle control and ground checkout, time between overhauls measured in hours, and operational dependability must be coupled with high performance to meet the propulsion needs of a space transport system.

The concept expressed in these words eventually became the foundation for the Space Shuttle engine. It was hoped that the thought of long life and reuse would be associated with Pratt & Whitney.

Figure 5.7 is a physical comparison of the J2 Apollo engine with the RL20, illustrating the concept of high pressure. The text of the brochure continues by contrasting the physical size of the 200,000 pounds of vacuum thrust of the J2 with the 250,000 pounds of sea-level thrust of the RL20. The J2 operated only in space and had an expansion ratio of 27.5. The RL20 would produce 265,000 pounds of vacuum thrust under similar conditions, albeit at a slightly larger diameter than the sea-level RL20.

Figure 5.7 Comparison of the RL20 and J2 engines.
(Courtesy of Pratt & Whitney)

The potential benefits to be obtained from operating at high pressure were described and illustrated by the ideal performance curve for oxygen and hydrogen shown in the brochure as Fig. 5.8. The optimum specific impulse and the ratio of oxygen to hydrogen at which it occurs both increase with increased area ratio in the exhaust nozzle, which is made possible by higher chamber pressure. In addition to the fuel consumption improvement, the increase in oxygen relative to the lighter-weight hydrogen increases the bulk density and permits a smaller and lighter vehicle to lift the same weight. Considerable interest had been generated in the space

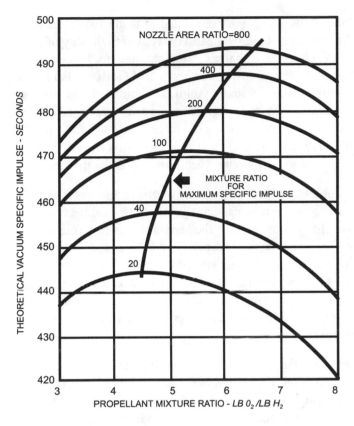

Figure 5.8 Ideal oxygen/hydrogen performance.
(Courtesy of Pratt & Whitney)

contractor community about the details of such a power plant, and the number of requests for data packages grew quickly.

Until this time, the Air Force and Pratt & Whitney had supported all of the experimental work. In 1965, NASA began to show interest in the program, and a 350K hydrogen turbopump project was launched at Pratt & Whitney in September under contract to the Marshal Space Flight Center. I was an assistant chief engineer of the project, and Bob Atherton was development engineer. Frank Williams was assigned to us as project engineer for the NASA pump programs. Frank had been responsible for the monster heat exchanger and the first engine in the 304 program, and later he played a major role in the RL10. (See Chapters 3 and 4) We were thrilled to have old "Steady Eddie" return for this very difficult job. This project was to be our first attempt at lightweight hardware, and the old metabolic leprechaun was right there, ready to pounce. Frank again claimed that considerable aging occurred on this program; however, the handsome young man in the photograph (Fig. 5.13 later in this chapter) with the pumps disproves this claim. I think that the intense technical challenge was something we expected in this endeavor, and everyone involved enjoyed it.

The first test of the 350K pump was scheduled to begin two weeks before my vacation in 1965. The gremlins arranged to have the test delayed until the day of our departure. We were to sail from Palm Beach to the Bahamas that night, and my loyal crew of Don Riccardi, my wife Carol, and our kids were stowing the boat. Bill Gorton, also a sailor, made me understand that I was not to leave for vacation until we completed that first test. The day dragged on. Finally, around 9:00 P.M. that night, the test was attempted. It was aborted by some failure of the test hardware that would require at least two weeks to fix. My decision was that the first test had been completed, even if the result was negative. Our kids were already asleep in the forward cabin of the boat when I arrived at the marina.

One of the first troubles emerged with a bang on the second test. A high-strength aluminum alloy normally used for aircraft wing spars, AMS 4135, had been chosen for the turbopump cases to reduce weight. Unfortunately, the alloy also has low fracture toughness, which was a new phrase in our vocabulary. The pump style is called a double volute type and was selected to balance the forces on either side of the case. The internal loads in a high-pressure pump are enormous—approximately 600,000 pounds trying to push the front housing forward. The fluid dynamic requirements define the shape of the cutwaters, one on each side, to accept the flow exiting the rotor in each stage. The sharper the edge was, the better from a performance viewpoint.

On the second test with the pump mounted on the test stand, Jim Sandy, Arno Kalb, Bob Davis, designers, and some others, whose names I have lost but who had worked on the design of the pump, wanted to see the test. However, the control room was too crowded. As usual, the critical tests were being performed at night, and those interested climbed onto the roof of a support building approximately 200 feet away from the test site. They thought this would be a safe distance for watching the test. As the pump accelerated part way up to speed, a loud boom and a fireball occurred, and the open sheet metal roof of the stand blew straight upward. Arno looked up and saw the shower of falling roof pieces, and he decided it was be a good time to climb down the ladder. Everybody but Jim had already departed. Jim says, "The redhead [Arno] took off, and he was faster than I was. I can remember running behind him, slipping and sliding on the stuff he was dropping." Jim also remembers hearing the shower as the things that had been blown skyward now rained down onto the roof. Everyone safely descended the ladder to the ground but kept on running, fortunately unscathed except for perhaps a problem the laundry might have noticed.

When the case is pressurized, the case itself and the bolts that hold the assembly together stretch a small but finite amount. This same stretch must be accepted by the cutwaters. With the high-strength material, such a stretch evidently exceeded the ultimate strength of the material, particularly with the sharp edges of the integral cutwaters. What was needed was an alloy with a lower ultimate strength, which could yield locally in the cutwater area without fracturing. Adequate strength was present in the balance of the material to hold the pump together if the cutwater could be relieved. Dr. John Mertz from the metallurgy department in East Hartford recommended a change to AMS 4130, a more forgiving alloy with yield and ultimate strengths a bit farther apart. The housing was cut back at the cutwater and given a large radius. An insert that carried no structural load satisfied the sharp edge requirement for fluid dynamic purposes. One of the things learned from this failure was that cryogenic temperatures tended to accelerate crack growth. A high-cycle fatigue failure from an undetectable crack could be made to occur in many fewer cycles. This led to a routine spin test of F100 fan disks at liquid nitrogen temperature at each overhaul period. Successfully passing this test proved that the disk had sufficient life to operate until the next inspection.

In early testing, significant vibration was exhibited, and the balance of the large opposing forces on the rotor tended to swap direction during acceleration. Two modes of vibration occurred at different speeds. In his spare time, Bob Davis in analytical design constructed an early electrical model of the rotor/support system. It was part analog and part digital and could simulate the two modes—one bounce, and one rock. Bob's work with the model showed that the radial stiffness in the bearing supports was too low and was the major cause of the problem. In today's world of fancy software models that can simulate almost anything, this may not seem like an important discovery. However, in 1965, it was a breakthrough.

I distinctly remember the meeting in my office, and everyone in the group was present when Bob Davis presented his findings. His simulation suggested the stiffness of the bearing supports of the shaft should be increased significantly to push the frequency up out of the operating regime. At that time, ball bearings were used, which are limited to a maximum spring rate of approximately one million pounds per inch. If we could change to roller bearings, the spring rate could be increased to three million pounds per inch, and Bob thought that was enough to solve the problem. The ball bearings originally were believed be necessary to take the residual axial load from the thrust balance system. From that point, an avalanche of ideas emerged. What if we changed to a double-acting thrust piston that worked in both directions, and tie the position of the shaft down so we knew where it was at all times?

The thrust balance piston essentially became two hydrostatic bearings, face to face. We needed a material to line the pressure plates in the housing so the piston could bump on startup before the gas pressure was high enough to centralize its position. A leaded bronze called B10 was selected and worked from the first try. It had been used to solve a dry bearing problem on the monster transporter used to move vehicles from the assembly building to the launch pads at Cape Canaveral.

Stiff roller bearings now could be used on both ends of the shaft. However, roller bearings previously were guided by the rail on the inner race, and they needed significant internal clearance. Internal clearance is an anathema to these high-speed applications because the bearings tend to skid on acceleration and the effective spring rate is reduced. We needed a new roller bearing with no internal clearance and a guidance system that did not depend on touching the rails. Thus, it was invented. A flat belt remains on its pulley without flanges because the pulley has a crown across the face. The belt always tends to climb to the high side. We wanted a zero clearance bearing during start, so why not shrink a thin outer race over crowned rollers, and open the end clearance between the rails on the inner race so they would not contact the rollers if they wobble? Extensive bearing rig testing proved the validity of the concept. Several bearing configurations were tested, the longest for nine hours. Bill Creslein and I have a patent on this bearing, as shown in Fig. 5.9. Bill was the APE under

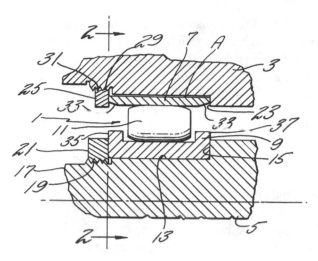

Figure 5.9   U.S. Patent #3,628,836 drawing for the new roller bearing.

Frank for the fuel turbopump. Our group was extremely inventive, and this particular event revolutionized our approach to pumps. Boost glide, which was the next new design and is described here in Chapter 6, incorporated all of these ideas from the outset.

The fuel turbopump rotor (Fig. 5.10) had a screw type-inducer and two unshrouded centrifugal stages, to create the 5500 psi needed to support a chamber pressure of approximately 3000 psia.

Around this time, a new RL20 mockup was constructed to illustrate the two-position nozzle (Fig. 5.11), and it was displayed at the exhibition of the Air Force Association annual meeting exhibition in Washington, DC, in September 1996. This concept originally was invented by Lou Billman at the United Aircraft Research Laboratory (UARL) in 1948 when he saw a collapsible Boy Scout drinking cup. High area ratio exhaust nozzles in the upper stages become quite long and often are limited by the allowable interstage length rather than diameter. By slicing the nozzle at a

Figure 5.10 The 350K hydrogen turbopump rotor.   (Courtesy of Pratt & Whitney)

Figure 5.11 Mockup of the two-position nozzle.
(Courtesy of Pratt & Whitney)

station that is large enough in diameter to be pulled up over the front of the engine, the length can be reduced by 40%. After stage separation, the nozzle skirt is deployed by three jack-screws. Such a nozzle was part of the XLR129 engine design and was tested in the 250K size. (See Chapter 6) The concept also has been flown in two models of the RL10—the A4 with an uncooled columbium skirt with an expansion ratio of 55 raised I_{sp} to 449 seconds, and the A4-1 with an expansion ratio of 59 raised I_{sp} to 451 seconds. The trick is to add the effective expansion area at low weight. The skirt on the B2 is made of ultra lightweight carbon/carbon imported from France with an expansion ratio of 285 to 1 and an I_{sp} of 466.5 seconds. For sailors, it can be likened to setting an ultra lightweight spinnaker.

Vice President Hubert Humphrey visited the U.S. Air Force exhibit where the mockup was on display. It was set up at the front end of a small room designed for presentations, and there were six rows of folding chairs, with aisles on each side. Dick Coar and I were standing in the back of the room when Vice President Humphrey entered from a side door and approached the mockup. I was to greet him and explain how it worked. The seats were empty, but both aisles were blocked with people standing. Wishing not to keep our guest waiting and because I have long legs, I stepped over the folding chairs and proceeded toward Vice President Humphrey. Dick Coar told me later that I was so focused on our important visitor that I did not see the two Secret Service men converging on me from both sides. They evidently concluded that I was harmless and allowed me to proceed. Vice President Humphrey was very impressed when the retractable skirt was demonstrated and asked if we had told NASA about our work.

Figure 5.12 shows the test history curve, which plots pump discharge pressure achieved versus time. This curve tells the grim story of the long and difficult turbopump development process. More than two years were required to achieve the design point of 5500 psi. Each symbol represents some significant modification in design. A six-month wait between points, which yielded little increase in pressure, required great persistence. Sometimes, the particular problem would end in a failure where the fragments would fill a barrel with pieces the size of walnuts. The peak pressure finally achieved in this program was 5862 psi. Loren Gross was the program manager for the turbopump programs at NASA Marshal, and Frank remembers him as always being ready to help.

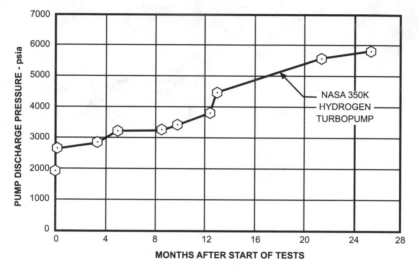

Figure 5.12 The 350K hydrogen turbopump history.
(Courtesy of Pratt & Whitney)

Del Tischler, who was head of propulsion technology (OAST) in NASA headquarters in Washington at the time, followed this work closely. He called Frank to congratulate him on having exceeded the target of 5500 psi and for having the perseverance to continue in the face of many obstacles. The next challenge was the companion 350K oxygen turbopump that benefited from lessons learned during the fuel pump program. Figure 5.13 shows Frank with the two pumps.

The rotor for the liquid oxygen turbopump (Fig. 5.14) was much simpler, having an axial inducer and a single shrouded impeller. Hank Vaughn was Frank's APE for the liquid oxygen turbopump. The single-acting thrust piston was spaced a significant distance from the turbine to provide ample space for the shaft sealing mechanism, which was borrowed from the RL10 design to ensure that the oxygen remained separated from the hydrogen. As in the RL10, initial tests were performed on liquid nitrogen substituted for liquid oxygen. The power to

Figure 5.13 Frank Williams stands behind the 350K fuel (left) and oxidizer (right) turbopumps. (Courtesy of Pratt & Whitney)

drive this rig was supplied by blowdown through the turbine of hydrogen gas from tube trailers. This turbopump was successful almost from the outset and achieved a peak pressure rise of 5601 psi. Both 350K turbopump programs were completed in June 1967.

The high-pressure engine work was to continue at an increased pace, as covered in Chapter 6.

Figure 5.14 The 350K liquid oxygen turbopump rotor. (Courtesy of Pratt & Whitney)

Boost Glide and the XLR129— Mach 20 at 200,000 Feet

In mid-1965, Mr. John Parangoski visited Bill Gorton to discuss a worry that was growing in the Central Intelligence Agency about a possible vulnerability of the A12. Previously, it had been thought that with its Mach 3+ speed over a defended site, the aircraft would be past the site before it could be hit. After detection by radar, no time existed for the decision, identification, command, and missile flight to occur. The concern was that the radar system would be "netted" so that the early warning stations could detect the incoming plane and pass the information to downstream batteries that would have time to respond. I never knew John's correct title in the CIA. However, he was in the office responsible for the equipment for airborne reconnaissance: the U2, the A11, the A12 (later the SR71 under the U.S. Air Force), and other aircraft. The Suntan (Chapter 3) had been under the U.S. Air Force. Ben Rich in his Skunk Works claims that the Air Force had jumped onto the hydrogen bandwagon because it had missed its chance with the U2.

John told Bill that McDonnell Aircraft in St. Louis had made a preliminary proposal of a system that was faster and might solve the problem, but it needed a rocket engine. Therefore, John wondered if we had anything suitable? With the ongoing high-pressure technology program, the timing was right for Pratt & Whitney. The first step was for me to go to St. Louis and meet with a man named Bob Belt. I thought I must have done something wrong because the McDonnell Aircraft security personnel were suspicious. They would not allow me to see Bob, and I returned home. Bill reported the outcome to John, who admitted that McDonnell had made an error. I returned to St. Louis with the proper introduction, and I met Bob Belt and Harold Altus. Both men were outstanding characters and people after my own heart. Another top secret program began for our group.

McDonnell Aircraft was proposing a manned glider to be launched by a rocket engine to Mach 20 at 200,000 feet. With a hypersonic L/D of approximately 3.2, which the company had confirmed through wind-tunnel and scale-model free-flight tests, the glider would have a range straight away of 4,800 miles and approximately 3,500 miles in a curved path to either side. Approximately two-thirds of the weight was carried by centrifugal force, and approximately one-third was carried by aerodynamic lift. The glider was almost skipping across the top of the atmosphere. It was to be dropped from the wing of a B52 at an altitude of 35,000 feet, and from two takeoff airports, it could cover 96% of the surface of the earth. The remaining 4% was over the ocean and of no interest. Having passed over Russia and China, the glider would leave the Asian landmass, doing faster than Mach 12.

Provision was made for loading the propellants from the B52 after its climb to altitude. The weight-carrying capability of a B52 from a wing pylon was limited to approximately 167,000 pounds. Using oxygen/hydrogen propellants, McDonnell staff had calculated a required glider weight of 132,000 pounds and would need an engine with 225,000 pounds of thrust at 35,000 feet.

Satellites already were providing reconnaissance from space and avoided the potential of pilot loss, which was a politically appealing asset. Because the time of flight was so predictable, however, targets increasingly were moved or hidden at the time of passing. Clouds often covered the area of interest at a critical time. Waiting was required until the satellite was in position. What the reconnaissance community wanted was an on-demand system that could be deployed at a critical time.

The McDonnell Aircraft proposal seemed to fit the bill for the reconnaissance community, and it was decided to proceed as soon as the money could be identified. At Pratt & Whitney, the preliminary engine design was already underway, and "old man Mac," as the chairman of the corporation was called, would keep the vehicle going. This was to be a cooperative venture by all of the reconnaissance community. To start the secretive engine development with an excellent cover, the Air Force would begin a demonstrator program in this thrust class at the Edwards Rocket Laboratory. The laboratory had been supporting rocket technology in components of a smaller size, and a competition for a larger demonstrator would seem normal. However, it was not a true competition. Rather, it was the start of development of the engine called the XLR129 for the McDonnell vehicle. If I ever heard a name for the program, I was brainwashed to forget it, and the name is not in the file. This was another one of those programs known by only a few people. John Foster, the director of Defense Research and Engineering, was a key player. However, one of his lieutenants in the chain for rocket development programs was not cleared and did not know the real purpose of the project. Harold Schultz, a kindly white-haired individual who was almost completely blind, questioned why the thrust of a demonstrator engine should be so finely set to 225,000 pounds and changed the RFP to read 250,000. We were asked if the change was acceptable, and we replied (with only a slight increase in pitch in our voices) that the change was fine.

Bob Belt of McDonnell Aircraft was a great guy with enormous technical capability. Bob loved golf, particularly the East course at MacArthur's old PGA course in Palm Beach Gardens in Florida. He also was the recipient of an early heart valve transplant. It was of the variety of a ping-pong ball in a wire cage, and Bob's only complaint was that he could hear it click at night when he laid in a certain position.

The real challenge was to build a hydrogen/oxygen bird with a structure and insulation that were light enough to meet the performance requirements. I remember that the target at McDonnell Aircraft was 4 pounds per square foot including insulation, which was a very low number. With a technology that boggled the mind of an engine type, it seemed as if Bob and his team might succeed. Look at the picture of the sample shown in Fig. 6.1.

Figure 6.1   McDonnell titanium structure. (Courtesy of John Robson)

The search for a lightweight structure suitable for the hydrogen/oxygen bird led to the invention of roll bonded titanium. In addition to its strength and light weight, titanium is one of the easiest metals to diffusion bond. If heat and pressure are applied to two pieces of the metal in intimate contact, a molecular migration occurs between the two and they become one. The boundary between the two pieces disappears. The sample in Fig. 6.1 has a slight curvature, which could be part of the process if desired and will be discussed later. The best way to describe the structure is to cover the manufacturing process step by step.

A flat sheet of titanium about one-eighth-inch thick was placed on a slightly larger sheet of one-half-inch thick steel. All of the pieces in this buildup were scrupulously clean. A series of parallel rods of pure iron with a triangular shape, almost touching, were laid lengthwise over the titanium sheet. Preformed strips of 0.040-inch thick titanium sheet in the form of inverted V's were hung over each of the iron rods. Diamond-shaped iron rods then were laid into each of the valleys between the titanium strips. Another set of the 0.040-inch thick titanium sheet V's—this time right side up—were laid point down into the valleys formed by the diamond-shaped rods. The remaining valleys were filled with a second set of triangular iron rods to fill the stack. A one-eighth-inch thick sheet of titanium—the same size as the first—was laid on top of the pack, and that was covered with another sheet of one-half-inch thick steel, the same as on the bottom. With suitable filler strips at the edges, the top and bottom steel sheets then were welded together. The complete assembly was evacuated and sealed.

This package was sent to a rolling mill, where it was processed similarly to the processing of any other piece of three-inch thick steel. The slab went through the heating and rolling sequence until it was reduced to approximately one-third of its original thickness. No one at the steel mill was any the wiser about what was inside. The one-inch thick "steel" sheet was returned to McDonnell Aircraft, and the steel was cut, leaving all of the titanium pieces diffusion bonded into one piece with iron rods in the core. The titanium did not stick to the iron or steel. The package was soaked in a nitric acid bath until the iron rods could be pulled out of it. What remained was a high-quality titanium structure with perfect diffusion bonds and a weight of only 1.4 pounds per square foot. If a round or tapered piece was desired, the "steel" sheet could be rolled at the steel mill prior to delivery. Those guys at McDonnell Aircraft in St. Louis were no slouches!

The parallels between this "competition" and the Space Shuttle Main Engine (SSME), where the winner was decided in advance, are ironic. I can remember boarding an airplane in California in the old-fashioned way that existed before jetports, where the practice was to line up on the field and climb up the rollaway boarding stairs into the plane. Two engineering types in front of me were having a heated argument about some problems they were having with the manufacture of a torroidal thrust chamber. With great restraint, I suppressed the urge to tell them it did not make any difference.

At Pratt & Whitney, we had always thought that the Aerospike concept did not make much sense. To stretch out the most difficult part of a rocket thrust chamber and have no direct structural path to hold it together courted trouble. To function as advertised required unrestricted flow of secondary air inward from the outside to provide the supposed altitude compensation. If the engines all were clustered in a group—particularly if they were gimbaled—each engine would see the wake of its partners for part of its circumference, and nobody knows where the thrust vector would be. The big difference between the two competitions was that we took nothing from them.

Until this point, the high-pressure staged combustion component work had been a vigorous program that sought to establish a sound database in this new technology for all of the components. The approach was logical, step-by-step, and proceeded at a pace that could be covered by the available resources. The engine development would occur sometime in the future. Suddenly, that future was now. Although few people were aware of the reason, the target was to develop the best engine we could as quickly as possible. Focus was placed on a very specific objective.

As soon as the requirements were sketched with Bob Belt and Harold Altus, the hard design of the engine began. In November 1965, the company began a major investment in the test facilities that would be required for this engine. The philosophy that it was up to United Aircraft Corporation to provide all of the things needed to do the work was motivated partially by a desire to avoid government dictation, as well as a genetic disposition from the old days. The stand was called E8 and, in addition to the capability of testing the complete engine, required a system of high-pressure tanks that could deliver the propellants in both gas and liquid forms, to test components individually. The most complicated stand ever attempted, it cost more than eight million dollars in 1965 dollars.

We did not recognize at the outset that the development of this facility would be almost as difficult as the development of the engine. The equipment suppliers to the chemical and oil industries provided many of the hardware answers. One of the first questions asked was: What kind of joints should be used for the stainless steel pipes that carried hydrogen gas at pressures of 10,000 psi and cryogenic liquids at levels almost as high? Fortunately, the oil industry had developed a clamped joint, called a Grayloc, that filled the bill and was suitable for field installation. Some of these pipes had walls that were two inches thick and weighed 250 pounds per lineal foot.

The tanks to contain the propellants were a special case in themselves. Twenty-four-hundred gallons of liquid hydrogen at pressures up to 6600 psi were required for test duration of ten seconds for an XLR129 engine component. High pressure dictated a spherical shape, and this one had an internal diameter of eight and one-half feet, with stainless steel walls that were eleven and one-half inches thick.

Etched in my memory is a supplier to the chemical industry in Alhambra, California, which took on the construction, as well as my trip to see the project underway. The individual gores shaped similar to a sixth of an orange skin each weighed approximately six tons. They were cut from flat stock on a machine, similar to a barrel stave maker, that was programmed to produce the correct bevel at each position. This would ensure that when properly shaped, they would fit together with a uniform weld gap. On the day of our visit, an elderly man on whom the shop was totally dependant was forming the gores. A large hydraulic press sat in the center of a room that had a sand foundry floor. The anvil had a rounded top that resembled a large mushroom, and the hammer had a corresponding convex shape. In a furnace on the side, pieces were heated while supported on a spindle. The operator had a control in his hand that could raise, lower, translate, and rotate the red-hot piece in any direction as he moved it back and forth between the furnace and the press. The scene was as if a ballet was being performed in Dante's inferno. Totally by eye and feel, this master would press the piece, which gradually assumed the correct shape, until he had six pieces that fit together with unbelievable precision.

The Russians had developed a welding system called "Electro Slag" to join thick armor plate in one pass. Formerly, large plates had required multiple passes with heavy chipping between passes to remove any residual slag and impurities. With the Electro Slag system, the eleven and one-half-inch pieces could be welded in one pass. Two gores were mounted on a fixture that allowed the parts to rotate around the geometric center of the tank, allowing the joint to move vertically between two shoes fixed at the equator. The copper shoes, which were insulated from the tank, conducted the welding current through the slag, which became the heating element. An inclusion-free weld was produced in one pass. Each of the six gores was attached in the same way until the complete sphere was produced with a manhole top and bottom. This core was suspended on long tangential shrouds within the outer vacuum jacket that included radiation shields. Filling of the forty-ton tank required five days, using liquid nitrogen as a prechill to cool the core down to liquid hydrogen temperature. Moisture contamination, which occurred once as a result of barometric changes over an unattended period, required ten days to warm the tank for cleaning. The liquid oxygen tank was the same except smaller—approximately 900 gallons.

Mixers were required to put the liquid and gas phases together so the correct inlet conditions could be simulated for components throughout the engine. Three different sizes of control valves were set up in parallel for each propellant to cover the wide turndown ratio. Feedback was included for all of these systems to ensure that propellant conditions could be maintained accurately throughout a test. Sophisticated advance and abort systems were set up to minimize damage during a malfunction. Fourteen months were required to construct and check this facility for the first preburner tests. Another three years of testing with the engine hardware were required before the operation became repeatable. Even then, constant vigilance was necessary to keep the thousands of valves and the control system performing properly in sequence at the correct moment. This incomparable facility allowed Pratt & Whitney to do things no one else had ever done.

Testing began on the E8 test stand in March 1967, when Dan Sims and Hal Gibson began with the variable area preburner. Operation was needed over a wide range of thrust levels, and the hope was that the engine could be throttled through direct variation of the hydrogen orifices in the preburner. Dave Bogue, who was in the design group at the time, came up with a clever mechanical design that worked well off the stand but locked up when in operation. The problem was that the injector bulkhead would deflect under pressure, and it was impossible to move the mechanism. It became necessary to preset a particular level and then run the test. The major problem was that it produced a non-uniform temperature profile that varied over the thrust range, and the close-coupled turbines—only 23 inches away—could not accept that. After 104 tests, its performance remained unsatisfactory for use with the turbines but was suitable for the staged combustion test with the radial spray-bar main injector, the transpiration-cooled chamber, and a two-position nozzle—all in the 250K size.

Taming the E8 Goliath—with its high-pressure liquid and gas propellant tanks and its three-stage parallel control valves that covered the wide turndown ratio—required a frustrating four months of agony, according to Hal Gibson. After several weeks of "resting" night after night on the lunchroom tables, Hal decided that he and Dan deserved something better. He bought a 1938 DeSoto that had an enormous rear seat, big enough to accommodate Dan. Hal slept across the front seat, and Dan, with his abnormally long legs, occupied the rear. Invariably, when awakened by the test crew for the next test, Dan's long legs were draped over the back of the front seat, often resting on Hal's head. However, the arrangement was more comfortable than the lunchroom tables.

Norm Bott remembers standing with another observer a distance behind the test stand and using binoculars to check that the igniter was firing properly. This test was to be limited solely to the igniter; if it worked properly, the blue light could be seen. The blue flash was observed and the high-pressure hydrogen tank coincidentally vented down. Venting releases a large quantity of hydrogen gas, which flows out through a 200-foot-high burn stack approximately 4 inches in diameter. This produces a loud roar and a torch of the burning gas 500 to 600 feet high. The observers had not expected this event, and they ran off as rapidly as possible. Norm was first in running, and he remembers a set of size-16 feet running behind him and coming down on top of him. He thinks it was Dan Sims, and the size 16 shoe would fit that scenario; however, Dan does not remember that. Whoever he was, that person had big feet.

These historic firings, of which there were 26 in the period of August through September 1967, had a total duration of 441 seconds. The two-position nozzle skirt was deployed after rated chamber pressure was reached and demonstrated its operation with three synchronized, recirculating ball jacks. Figure 6.2 shows the spectacular shock diamond pattern, which was produced by a nozzle expansion ratio of approximately 40/1 at the 186/1 pressure ratio between the main chamber and the atmosphere. The Mach number was approximately 6. The transpiration-cooled chamber (Fig. 6.3) was developed from the early 10K experimental work described in Chapter 5 and was converted to a flight weight design. The coolant flow was less than 1% of the hydrogen propellant at rated thrust. Transpiration-cooling effectiveness suitable for unlimited reuse was demonstrated. Thermal strains associated with convective cooling were avoided.

John Chamberlain provided the input for every combustion and cooling aspect of the engine. He was on the stand many nights during the struggle to make it all perform. John has always enjoyed working with hydrogen and says it does everything better than any other fuel. He tells about an example of the wide combustion limits of hydrogen. Once, when running a preburner test with cold propellants, the liquid oxygen flow was reduced to determine the

Figure 6.2 Night firing of the 250K staged combustion. (Courtesy of Pratt & Whitney)

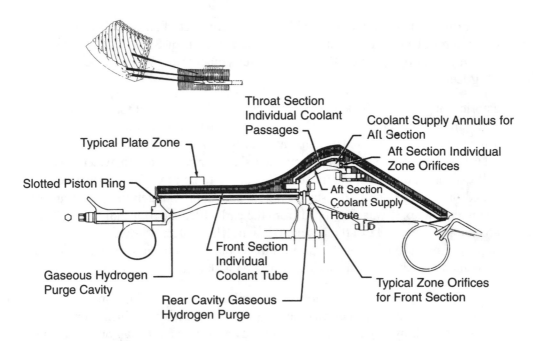

Figure 6.3 The 250K transpiration-cooled chamber.
(Courtesy of Pratt & Whitney)

blowout limit. The combustion temperature dropped below 32°F, at which point ice built up inside the burner and "snow" poured from the nozzle.

In these tests, an orifice simulated the turbine pressure drop of 1500 psi. The main injector configuration evolved from work in the 50K size. A combination of tapered radial liquid oxygen spray-bars (Fig. 6.4) of three different lengths were cantilevered from the outer manifold. The injector spigots, each of which had internal swirl ribbons, projected through the corresponding slots in the Rigi-Mesh face. The hot hydrogen-rich turbine exhaust gas issued through these slots, and the cantilevered support of the cold spray-bars, with a sliding inner constraint to maintain slot alignment, avoided thermal distortion. The superior mechanics of this design were coupled with excellent combustion efficiency.

NASA had encountered serious combustion stability problems in some of its earlier engines, and a special department was set up at Marshal that had an almost religious zeal. This department was charged with finding and stamping out any combustion stability problem, even in designs where instability did not exist. Although Pratt & Whitney had never encountered combustion instability with hot hydrogen gas systems, a series of six bomb tests with pulse guns (for a total of another 136 seconds duration) was run at the end of September. Total firing time of the high-pressure staged combustion hardware was within three seconds of ten minutes—a long time at this elevated metabolic rate. The development of the XLR129 engine for the "spooks," as the reconnaissance community is known, was proceeding well.

Figure 6.4 Radial spray bar injector (a) upstream and (b) downstream.

An interesting aside in this "age of spin" has surfaced. The traditional gadget to suppress potential combustion instability favored by the cult was to erect six radial fences equally spaced around the injector face. This was done on the SSME main injector at Rocketdyne as an offering to the instability gods. The fences, which projected upward from the face some distance, required hydrogen film cooling in the 6000°F environment. A new engine model, designed to operate at a lower chamber pressure to ease continuing problems, was not called a reduced pressure engine. Rather, it was called the large throat engine because that sounded better. It was announced that the loss in I_{sp} that would have resulted from the reduction in expansion ratio with the larger throat was more than made up by eliminating the film-cooled fences that were found to be unnecessary. Spin is everywhere.

With the demise of the variable area preburner injector, a new design was needed quickly. By June 1968, a new preburner concept invented by Carl Kah was being tested and showed a greatly improved temperature profile. It featured fixed-area fuel side and two-stage liquid oxygen injector elements. The liquid oxygen elements had a small swirl entry for low flows and a large swirl slot for full flow in the same element. Control was provided by a slide valve designed by Phil Scott, which was integral with the injector that would open the small entry to provide stable combustion at low throttle settings and open the large slots as the throttle was advanced. This design proved to have the required uniform temperature profile and stability over the complete range of thrust. To make this injector work, a sliding shaft seal on the valve stem was needed to allow it to move freely and at the same time prevent the leakage of oxygen at 6000 psi. A stack of thin Kapton sheets—a tough plastic—with an undersized hole was forced down over the stem. A shaped washer prevented the dimple in the many plastic sheets from toggling, and the plastic provided multiple corners to wipe the stem. It was completely leakproof in tests of more than 1,000 cycles at maximum thrust pressure levels with liquid oxygen. Lou Emerson and Joe Henderson in his control group invented the concept.

Development testing of this new preburner combustion concept continued. The objective was to improve its performance, particularly the temperature profile, in preparation for the forthcoming combination with the fuel turbopump in the power head.

The design of the 250K hydrogen turbopump was third generation, following the 50K and the 350K. It incorporated all of the lessons learned during more than half a decade of trying, finding limits, and trying again. The design team was drawn from those who had participated in the earlier pump designs. The secret to the rapid growth was the opportunity to return to the drawing board unencumbered and to correct fundamental difficulties encountered with a previous design. The two-stage unshrouded impeller configuration with a two-stage turbine was retained from the 350K design. The double-acting thrust piston, coupled with the special roller bearings, was incorporated. Figure 6.5 shows the 250K turbopump rotor.

Figure 6.5 The 250K turbopump rotor.
(Courtesy of Pratt & Whitney)

Significant changes were made in the choice of materials, most notably in the pump housings and the turbine blades. In place of the high-strength aluminum used in the 350K pump, the 250K pump used INCO 718, a tough nickel-based material that has outstanding properties for cryogenic pump housings. Continuing work by the Materials Laboratory found a hydrogen embrittlement problem with long-term exposure at elevated temperatures. Joe Moore and his staff developed an alternative based on A286, an iron-based alloy, offshoot of the German turbine disc material, Tinidur, for the turbine housings. The turbine adopted tough, directionally solidified blades, first introduced in commercial engines. These blade castings were grown in a special process that produced a series of small parallel grains from root to tip and resembled the pattern of a Damascus barrel. Compared to a jet, the turbine inlet temperature was modest in the rocket engine; however, the directionally solidified (DS) offered significant benefit in thermal shock resistance, which will be discussed in Chapter 9.

Figure 6.6 shows the project engineer, Bill Creslein, with the 250K and 350K turbopumps. Bill was the APE for the 350K turbopumps.

Figure 6.6 The 250K (right) and 350K (left) fuel turbopumps.
(Courtesy of Pratt & Whitney)

Driving the fuel turbopump, which required 55,000 horsepower, was done with a special bootstrap rig invented by Bob Atherton. Bob's strong hand and creative mind were behind everything we did, similar to those of John Chamberlain. Hydrogen from the pump discharge in a liquid/gaseous phase called "ISH" was ducted to a large tubular heat exchanger. Because the unit was more than 20 feet long, a special crane was required to lift it into the facility. To provide the power, steam from accumulators was directed through the outside of the heat exchanger, and the warmed hydrogen gas working fluid was expanded through the pump's own turbine. Marv Glickstein in analysis did the heat transfer design for the system.

Thirteen trouble-free tests, for a total of 405 seconds, attained a pressure rise of 6705 psi and qualified that the rig was ready for hot turbine testing (Fig. 6.7). The difference between the times required to reach design performance for each of the turbopumps was the measure of the benefits of having trod the path with the same people many times. The test history of the XLR129 shows an almost vertical rise to 5600-psia pump discharge pressure on the first build in two weeks. At that point, the rig was removed from the stand and disassembled, and all parts were inspected and found to be in original condition. The same parts were reassembled for the second build, and a discharge pressure of 6705 psia was achieved in less than five months.

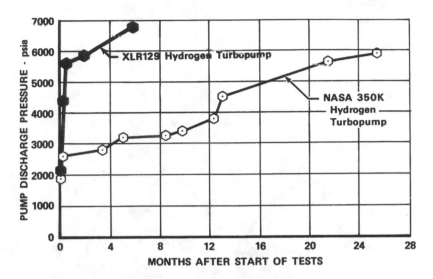

Figure 6.7 Turbopump history. (Courtesy of Pratt & Whitney)

On the E8 test stand, the power head was tested with the dual-orifice preburner injector and oxidizer valve with simulators for both turbines in six tests for 42 seconds, to qualify those parts for the next series, which would include the fuel turbopump. Figure 6.8 shows the complete assembly of the power head, except for the liquid oxygen turbopump, the absence of which will be covered in Chapter 8.

Bill Madden in the design group came up with the concept of tying all of the high-pressure components together with a series of spherical cases, which offers a lightweight and compact structure. The main case was in the center and resembled a heavy-duty diving helmet. At the intersection of each of the spheres, a heavy circular ring provided the hoop restraint that was removed when the spherical shells were pierced at the intersections. It also balanced any forces among the pieces. The case was strain gaged and passed all tests with flying colors. The hot gas and steam from the preburner passed through the small, transpiration-cooled ball in the center and ducted the flow into the turbines. The fuel turbopump on the right was plugged into the ball, as was the preburner on the left. The annular discharge from the turbine passed outside the ball down through the main injector (Fig. 6.4).

This elegant design was completely trouble free in all tests. It provided a solid mounting for all components and a straightforward flow path for the hot gases. All internal surfaces were transpiration cooled to ensure that ambient temperature or lower was maintained on all of the outer surfaces. In July and August 1970, this assembly was run in six tests for a total of 94 seconds (Fig. 6.9). During these tests, the required pump pressure rise of 5500 psi and a

*Figure 6.8 The complete assembly, showing the preburner, the main case,
and the fuel turbopump. (Courtesy of Pratt & Whitney)*

Figure 6.9 Hot turbopump test. (Courtesy of Pratt & Whitney)

turbine inlet temperature of 2411°F were attained. By this time, we had a dedicated team of engineers, technicians, and mechanics who had worked together on hydrogen projects for 15 years and on the high-pressure staged combustion rocket engine for a decade. The goal for which they had worked with such devotion evidently was at hand. Their experience had been focused specifically on solving the difficulties that were to be encountered in the development of this engine. The decision to take this job from them and give it instead to a chosen instrument was extremely costly to the country. The loss of this team and this elegant and straightforward engine concept was unfortunate.

The 250K hot turbopump testing was the last testing performed under contract with the U.S. Air Force. The McDonnell Aircraft concept was competing with something in the Air Force called the "Manned Orbiting Laboratory" and lost in a high-level decision. One of the truisms of working for the "spooks" was that programs could come and go without ever hearing an explanation as to why, at least at my level. You might have a "need to know" for your program but not for the one that displaced you. In the waning months, the contract had returned to a true demonstrator. The liquid oxygen turbopump was the next component in line. However, before it was funded, NASA had started the Space Shuttle campaign, and the Air Force gave the XLR129 program to NASA, granting free use of the existing hardware to Pratt & Whitney. NASA promptly canceled the liquid oxygen turbopump because it would be unfair to our competitors to fund it. I will bet there were times when NASA wished it had continued the program.

CHAPTER 7

XLD-1 Gas Dynamic Laser

In October 1967, I received a phone call from George McLafferty, one of my old friends from my United Aircraft Research Laboratory (UARL) days. George had done all of the super-sonic aerodynamics for the ramjet, as described in Chapter 1. Lately, he had been working in the field of lasers. George said, "Your rocket test facility would make a large gas dynamic laser." My response was, "What is a gas dynamic laser?"

Through the combustion of carbon monoxide with oxygen, an energetic population of carbon dioxide gas is produced, into which nitrogen and a small quantity of water vapor are added. The energy is split into three types in the molecule: (1) translational, (2) rotational, and (3) vibrational. The mixture is expanded rapidly through supersonic nozzles, causing the static temperature to fall abruptly. The first two forms of energy dissipate rapidly by speeding up the gas. The third form of energy hangs on for a while. To achieve the lower level required to be in equilibrium at the reduced temperature, the atoms throw off photons to get rid of the energy. If mirrors are arranged to face each other across the flowing gas stream, the photons will bounce back and forth between them in a straight line. If there is a hole in one of the mirrors, part of the photon stream will emerge as a high-energy laser beam. The scenario reminds me of the cartoon in which Wiley Coyote runs off a cliff while chasing the Roadrun-ner, and Wiley Coyote's feet keep running in mid-air for a moment before he drops to the canyon floor. This is a simple engineer's view of the essentials of a gas dynamic laser. It causes some indigestion in my scientific friends.

George and I talked for some time during this phone call. The following week, George came to FRDC. We hatched a plan to piggyback a stand directly south of the E8 stand that could share some of its equipment on a non-interference basis. Nothing could delay the high-pressure rocket work. I continued to watch the XLR129. However, Bob Atherton continued with his capable day-to-day management of the program, which freed me to look at this additional project. When George explained the basics of the gas dynamic laser, I said we could do the "plumbing and heating" part if UARL would supply the science.

Lasers are modern miracles that affect our lives every day: non-invasive eye surgery, compact discs, and scanners in supermarket check-out lines which read products and prices into the cash register. The list is endless. The power level of that type of laser is quite low, measured in millionths of watts. Here, we are talking of hundreds of thousands of watts. The potential of such a device as a defensive weapon against incoming missiles excited the U.S. Department of Defense, and the Advanced Research Project Agency (ARPA) funded research efforts through all of the services. The possibility of a weapon that had a muzzle velocity equal to the speed of light and that could be pointed quickly was a large incentive.

McLafferty and I approached our respective bosses with the plan, and we received the official approval to proceed in December. I think Dick Coar was on vacation at the time, and I remember going directly to Bill Gorton, who was enthusiastic about the project. Dick was rather cool to the idea when he returned, thinking our plate might already be too full to take on this diversion. However, as always, he was ready to help when we needed it. In typical FRDC fashion, the project was well on its way. All of the fluids needed, except for carbon monoxide, were already available on the E8 stand in large quantities. One of the gaseous oxygen tube trailers was converted to hold carbon monoxide. There was concern about the possible formation of iron carbonyls on the inside of the tanks, which might cause them to become brittle. However, periodic inspections never showed any signs of such a problem.

George knew that ARPA was interested in large lasers. Therefore, the nominal size of the XLD-l was set at 1 megawatt. It was estimated that producing this level of power would require a cavity approximately 22 feet long by 4 inches wide. The problem of how to fabricate accurate supersonic nozzles to supply this large active cavity led to the first invention of the XLD-1 program. The individual nozzles were two-dimensional and were formed by wafers approximately seven inches long that were precision cast of turbine blade material. Carl Kah remembers the rush to receive the castings from our vendors. They resembled linotype slugs. Each wafer, approximately one-eighth-inch thick, formed half of the throat and the curved nozzle expansion surface. The supersonic section was about one-quarter-inch long and rapidly dropped the static pressure to approximately 1 psi. The expansion ratio was 30/1. Figure 7.1 shows two of the wafers clamped together.

To maintain the height of the two-dimensional nozzle throat over the relatively wide four-inch span, lands were cast into the subsonic section leading up to the throat to control the spacing (Fig. 7.2). These lands eventually led to a problem with wakes, which was recognized later. Approximately 1,700 of these wafers formed the nozzle deck at the bottom of the laser cavity. The flow from the nozzles is upward.

Figure 7.1 The XLD-l wafer nozzles.

The laser cavity was a rectangular space approximately 4 inches wide, 1 foot high, and approximately 22 feet long. An 18-inch diameter tube of Hasteloy X that was 22 feet long ran under the nozzle deck and formed the distribution manifold from the burner to the nozzles. The tube had a slot cut into its top, to which a heavy flange was welded. The nozzle wafers were loaded into the flange. Tie bars were welded across from side to side beneath the nozzle deck to hold the tube together. The hot carbon diox-

Figure 7.2 Wafer nozzle lands.

ide with the nitrogen and water additives flowed upward through the nozzles into the cavity where the lasing occurred. The beam passed outward through the exit optical box at the output end. The spent flow continued upward through the vertical convergent/divergent diffuser and exited through the turning vanes. The white line in Fig. 7.3 traces the path of the laser cavity on the inside. The burner and its manifolds can be seen on the left end, slightly before the input box. The man on the right is reaching into the output box. The program was so active it seemed that the site constantly was under construction.

Figure 7.3 Side view of the XLD-1. (Courtesy of Pratt & Whitney)

The shed roof served the dual purposes of partial protection from the Florida sun and, more importantly, a security shield from prying satellites. Long cylindrical tanks stored gaseous oxygen used by both stands. Tube trailers contained carbon monoxide.

Originally, the plan was that UARL would supply all of the optics because it had been working with a variety of lasers for some time. The need for metal mirrors using rocket-cooling technology convinced us that Pratt & Whitney should build them. The first grinding and polishing machines were set up at FRDC under the stairs, in a tiny spot carved from the shop office. This part of the business grew as other people wanted to buy our mirrors, and a catalog was published. The work soon outstripped the space available, and the Optical Laboratory was moved to the rocket area. Some fancy technology was developed, which allowed the mirror surface to be manipulated to correct distortions in the beam. Eventually, a separate building was constructed, dedicated to the optical business. During the construction phase of the XLD-1, George McLafferty would sit on a stool located under an umbrella at the construction site and would field endless questions from the "plumbing and heating" side of the joint venture with Pratt & Whitney.

While the stand and rig were being constructed, far enough along to have convincing pictures of most of the pieces, George decided that we should visit ARPA and tell them what we were doing. Pratt & Whitney's work had been an all-private venture until that point, and we hoped to attract outside support for the long pull. George obtained a date for us to visit Lt. Col. John MacCallum, who was responsible for laser work in ARPA under Dr. David Mann. The colonel was working at his desk in the middle of a large room, and it was evident that our meeting was interrupting something of great importance to him. With the proper disclaimers about being sorry for the interruption, George said he thought John might be interested in some photographs of the work being done at Pratt & Whitney. As I took out the first photograph and George began to describe it, John immediately snapped to attention. By the time I had pulled out the third photograph, the colonel said, "Stop. I've got to go get Dr. Mann." Dr. Mann was equally impressed. Both men used the phrase "very impressed" many times and did seem amazed that we had built this big laser without their knowledge of the project. By the third week in June, we were under contract with ARPA through the Air Force Weapons Laboratory. We always were suspicious that the colonel had been working on sole source procurement paperwork for AVCO and Dr. Arthur Kantrowitz, who had done the lion's share of the gas dynamic laser work until that time. From that point, the credit was to be shared.

Testing of the XLD-1 began in mid-March, and a measurable output of 77 kilowatts was demonstrated in April. We were off and running but had a way to go to reach the rating. This early work was done as a simple oscillator configuration with a single spherical copper mirror at the front end and a similar mirror with a one-inch hole in the center at the output end. One of the early problems encountered was that stray oscillations would destroy the mirrors by hitting the edges off angle. On occasion when it was particularly bad, a telltale pile of melted copper mirror would be evident in one of the end boxes.

The early targets to see what we had were a variety of things at close range, the earliest right in the exit box. The first few were stacks of furnace insulating boards called Maronite, available locally from building supply vendors. The beam would penetrate the block to a depth of seven or eight inches and carve an inverse stalactite-shaped cavity approximately three inches in diameter on the first board down to a series of points in the last. I can remember Bill Gorton's expression when he looked into the cardboard carton we brought into his office at the end of the first week. The stack of boards could be separated to show the irregularly shaped holes that the beam had carved and, although denigrated by my scientific buddies as a soft target, was impressive to us beginners. The target was trimmed on the outside, fitted to a varnished mahogany box, and taken to Washington to show ARPA that the XLD-1 was real. The range was stretched to 50 feet, and various aluminum, titanium, and plastic targets demonstrated their vulnerability. One memory that I vividly recall is the hole vaporized in a granite block. Obviously, something very powerful was growing in the E9 facility.

Dr. Petras Avizonis was the chief civilian technician at the U.S. Air Force Weapons Laboratory for lasers. He paid his first visit to FRDC, and I can remember the trip to examine the E9 facility. For some reason, we took three cars. Ed Pinsley was first, I was second, and Pete was following me as we left the rocket area control room. The parking area for liquid oxygen trucks was located on the right side of the road, and Ed had already passed it. As I drove past the spot, a 12-foot alligator emerged from between the parked Dewars and proceeded to pass behind me with its typical stately step. I looked into the rearview mirror of my car and saw that Pete had stopped to let this monster pass. Pete is a true scientist, and his mind usually was focused on things other than his local environment. I swear I saw Pete's eyes dilate as the alligator continued its procession across the road in front of Pete's car and into the canal.

The United Aircraft Research Laboratory had been holding our hand, so to speak, with trips back and forth between East Hartford and FRDC through the first months. In June, Ed Pinsley was induced to move to Florida and became our resident scientist, much to my relief. Ed directed the scientific part of the project. We both worked on the engineering, and I continued the manufacturing and construction support. One of the major changes in direction implemented by Ed was a growing feeling at UARL that instead of an oscillator and its many problems, the XLD-1 could be better used as an amplifier of a controlled input beam from an electrical low-power laser oscillator. The laboratory provided the 100-watt electrical input laser. An optical bench and house were constructed to supply the input beam to the XLD-1.

This approach introduced another set of problems called feedback. The gain in the main cavity is very high, and if a small errant part of the cavity power is reflected back in the wrong direction, it can wipe out the input oscillator. In an effort to "harden" the input beam to these attacks, the laboratory later built a preamplifier to go between the input oscillator and the XLD-1, which raised the input power to 700 watts. Two quartz flats were added as angle reflectors to act as fuses in the event that feedback did occur.

In early September, the rig was returned to the shop, and seal strips were added and the nozzles were repaired to restore their flatness. By the middle of the month, the rig was reinstalled, and gain was higher than on the first build. The U.S. Air Force advised Pratt & Whitney to stop work because it had run out of money. On September 24, funding and work were resumed. Approximately 150 firings had occurred by this time, and the 100 kilowatt level had been demonstrated. By the end of the month, 150 kilowatts was reached.

On October 9, a demonstration test was made for Drs. Foster, Tucker, Rechtin, and Mann. In this test, a power level of 210 kilowatts was measured. Work was stopped on that day because funds again had been depleted.

We began to have a steady stream of scientific visitors coming to see this new machine. Many had been part of a secret group called the Jason Committee, which had been convened to study what could be done with a high-power gas dynamic laser. They had conceived of a preliminary design, and we learned later that their major cavity dimensions and gas compositions were close to those of the XLD-1.

To provide cover for this project, rocket terminology was substituted for laser language, such as laser power was expressed in pounds of thrust. The visitors were confused until given the code.

On November 4, a blue-ribbon panel of scientists assembled for a meeting on the "Eighth Card" program in Dr. John Foster's office. The name was the secret designation given to the laser work. The official attendees included:

- Dr. John S. Foster, director of Defense Research and Engineering for the Department of Defense
- Dr. Charles Townes, inventor of the maser, the predecessor to the laser
- Dr. O'Neill
- Dr. G. Millburn, consultant to IBM
- Dr. Herbst
- Dr. Eberhardt Rechtin, ARPA
- Dr. N. Kroll
- Dr. David Mann, ARPA
- Dr. Bloembergen
- Lt. Col. John MacCallum, ARPA
- Mr. E. Meyers
- Mr. George McLafferty, UAC Research Laboratory
- Mr. Richard Mulready, Pratt & Whitney
- Mr. J. Dempsey, AVCO
- Dr. Arthur Kantrowitz, AVCO

The discussions were wide ranging, and Dr. Foster wanted to get the show on the road. The question arose about whether there should be one or two 1-megawatt devices—one built by each of the contractors. Dr. Townes said, "to get the 1 megawatt more quickly in terms of dollars and time, the Pratt & Whitney device is the obvious choice." AVCO was not too happy, and Dr. Rechtin suggested that ARPA ask AVCO what AVCO could do with more money. Dr. Foster said, "OK, but let's not wait around any longer. Let's go." It seems as if George McLafferty was spot on with the bet on the 1-megawatt size.

I made the presentation for Pratt & Whitney with some trepidation in front of this august body. I can remember being asked by Dr. Millburn, who was quite negative about lasers, why we had built the XLD-1. What I wanted to say but did not dare was this: As a boy, I had been a duck hunter. Thus, it seemed to me that if you had a gun that could shoot ducks where they were, rather than where they were going to be, then singe off the pinfeathers, and cook them on the way down, you had a winner. Such a dark joke did not seem appropriate in these circumstances, so I mumbled something about our having a special facility. I am relieved to finally have an opportunity to tell that joke here, after 30 years.

Dr. Rechtin said that another suggested application was the use of a large laser beam from the ground to fire energy directly into the back end of a rocket. It would allow leaving a major part of the weight of a rocket system on the ground and would considerably improve performance. We had studied this possibility, and George noted that a rocket bell nozzle with a mirror surface was an almost perfect collector that would act similarly to a lens and focus the heat coming in the laser beam at the throat of the chamber. The oxygen system and tanks could be eliminated completely. The hydrogen gas, which normally is transparent, could be made to absorb the radiation by seeding the gas with various materials. Calculations indicated that the I_{sp} would be more than 2,000 seconds. The size turned out to be the killer. Rough numbers showed that a laser of 1 megawatt would be required to produce the thrust of 57 pounds—not much for a first stage. The Space Shuttle needs approximately eight million pounds of thrust at liftoff. There were a few other practical problems, such as maintaining perfect aim at the nozzle. It would take great courage to ride such a beam and hope that your buddies on the ground were good shots.

I remember a strange encounter that occurred after the meeting. Three of us were walking side by side down one of the broad hallways in the Pentagon. Dr. Kantrowitz was on one side, and I was on the other, and my memory now is gone regarding the person in the center. As the three of us walked along, a thought struck Dr. Kantrowitz, and he leaned forward and looked at me. He said, "You're the guy who stole the program away from us." I replied, "We took nothing from you, Dr. Kantrowitz, except for the limiting contraction ratio for a convergent diffuser that you had published when at the University. It was very useful to us for our ramjet."

In those days, one of the major wickets for lasers was to find a material suitable for the window to enable the radiation to leave the cavity. Most of the candidates were salt crystals of various types, such as gallium arsenide which was suitable for low power but was nothing close to what was expected to work for the high flux density we would have.

In a cab on the way to Washington, DC, the light bulb came on for George McLafferty for the invention that solved the problem. The neat part of his idea was its simplicity. The static pressure inside the laser cavity is 1 psi. If a hole were cut to let out the beam, outside air at 14.7 psi would tend to rush in at Mach 1, the amount in pounds per second set by the size of the hole. It happened that there existed on the E8 stand a large two-stage steam ejector system remaining from the RL10 early days (when the E8 stand was E3 and E4), which was designed to pull down to 1 psi. By arranging a circular pickup cavity inside the hole and extracting the inward-flowing air, it was possible to cut a hole in the endwall of the cavity through which the beam could pass into a region that effectively had a 1 psi static pressure. The steam-ejector-driven aerodynamic window was used on the XLD-1 throughout the program. One problem that did arise, particularly in Florida on occasions when humidity was high in the summer, was that the low pressure would cause water to condense and soak up 30% of the beam.

Later, Bud Hausman from UARL carried the idea one step further and devised a scheme that eliminated the need for steam ejectors. If the incoming gas to the window were taken from a high-pressure source and expanded along a curved path, at some point along that path, the static pressure would be 1 psi. If that location were made coincident with the end wall of the cavity, the hole could be cut to allow the beam to emerge with no inflow of outside air. For that invention, Bud received the annual United Aircraft Corporation award to an employee for the most important invention of the year. The award was named after George J. Mead, vice president of Pratt & Whitney Aircraft Company. This system also was used in the U.S. Air Force A11 program.

Bud and I had been roommates at a historic "bachelor pad" in an old mansion on Asylum Avenue in Hartford. Mrs. Southworth was the cook and housekeeper, and nine of us from various walks of life lived there comfortably. One day, the owner of the property concluded that he could make more money by converting the site to commercial use, and we learned that a zoning meeting was to be held on the subject. Although we did not think we had much of a chance, Bud and I decided to alert the neighbors that a meeting was to occur and to warn them that their quiet residential neighborhood might become commercially zoned. After dinner, we walked around the corner by a 10-foot-high manicured privet hedge and up a short side street to a beautiful house. The bronze plaque read, "Doctor." We rang the doorbell, and a young lady answered the door. We asked to see the doctor, and she invited us into the house. As we proceeded down a long foyer, we had a strange conversation. The situation resembled movies in which the questions and answers almost match but not quite. When the young woman asked if we were patients of the doctor, we told her that we were merely neighbors. "Isn't it nice of you to come?" she said. When we reached the living room, we turned left and there was the doctor, lying in a casket amid many flowers and candles. We quickly decided that the doctor did not care about the zoning meeting. We stood there for a respectable period and turned to leave. The young woman thanked us for coming and led us to the front door. Our big problem was to leave the area quickly and without looking at each other. We made it down the short street and hurried around the corner. Then, we fell into the hedge, in gales of silent laughter. We decided not to visit any more of our neighbors and returned to our quarters on Asylum Avenue to drink a toast to the recently departed doctor.

After our eviction from Asylum Avenue, George McLafferty and I lived in the successor in Glastonbury. We were golf partners in a league sponsored by the Research Laboratory. Neither of us was a great golfer, but I remember George's determination. He could hit the ball a mile, but it did not always travel a straight path. We were playing at Minnechog one day, and the particular hole was long and narrow, with a closely spaced grove of pine trees on each side of the fairway. George hit a tremendous drive that went straight for 100 yards and then turned toward the right. We could hear a sound similar to a mad xylophone as the ball bounced through the trees. However, George was not one to give up easily. When addressing the next ball, George cranked around slightly to the left to compensate. This time, the ball traveled farther before turning toward the right, but the same "plink-plonk" resulted. By this time, a red flush was beginning to appear on George's face, and he turned his feet almost 45° from the hole. On this third stroke, the ball went straight and made its music among the trees on the left side of the fairway. George did not swear much, but later that night as he closed our refrigerator door, I thought the door certainly would fall off its hinges.

Work on the XLD-1 resumed on November 18 with a letter contract, and Build 4 with the new nozzles was installed. Gain tests showed no major improvement over the original nozzles. Some cracking in the sharper trailing edges evidently had been caused by vibration. Build 6 was assembled with the old design nozzles from Build 3 and exhibited a substantial improvement in gain over Build 3 at heights of 1.5" and 4.0" above the nozzle deck shown in Fig. 7.3. However, we did not know why this occurred.

The beam path through the cavity was Z-shaped and tended to push the last pass upward into a lower gain region. We reasoned that the use of cylindrical mirrors would produce a rectangular-shaped input beam and allow the three passes to be pushed downward into the most active high-gain region. The design and manufacture of such optics was begun, and a special Barnes camera that could photograph the beam was ordered.

A most important factor in obtaining optimum conditions in the cavity is the operation of the convergent/divergent diffuser and its ability to hold down the test section static pressure to the proper level to match that produced by the nozzles. Data indicated that static pressure varied significantly throughout the cavity, and a redesigned diffuser was installed in January 1969. With the original diffuser and nozzle deck removed, three cracked tie-bars were found, a sign of their severe service. During the two and one-half week shutdown to install the diffuser, UARL staff continued to check the new cylindrical optics. They produced a flat elliptical beam three and one-quarter inches wide by seven-eighths inches high that would pack the three passes downward tightly into the high-gain region within the four-inch height above the nozzle deck. Build 6 was installed with 1,754 of the new nozzles tightly packed, cylindrical optics, and the new diffuser. The new diffuser had a slightly increased contraction ratio and exhibited some initial starting trouble. George McLafferty's perforation work from the ramjet days came to the rescue. The starting problem was solved, and the new diffuser dropped the cavity pressure to 1 psi, the lowest achieved yet and uniform throughout. The gain at the four-inch level had doubled to 104, compared to the previous 51.

As with all good things, the high gain brought with it some trouble. It made the XLD-1 act similarly to a self-powered oscillator, and it would throw powerful feedback beams back into the input system. It is a good thing the quartz fuses were in place. The theory was that if the hydrogen gas flow to the rig were delayed until the input shutter opened, the high gain might be delayed until the input beam was established and the feedback might be avoided. The gain was suppressed, and no feedback occurred until the hydrogen was turned. Bang! The gain jumped to its former level, and a strong feedback beam shot from the front end.

The high reflectivity of the sidewalls of the laser cavity at low angles caused part of the beam to skip from side to side. Anything—even a macadam road—is highly reflective at such a low angle. It was thought that the feedback beam was slightly off axis, and this idea led to a partial solution. Ed and I shared the invention of adding vertical fences approximately one-half inch high and long enough to span the cavity in the direction of flow, spaced to catch the skip and stop the errant beams. The input system was moved back to reduce the solid angle of view. These changes reduced the feedback but did not completely eliminate it. In February 1969, a power level of 210 kilowatts was repeated with a circular beam with minimum feedback damage to the input system. The next test changed back to the closely packed rectangular beams in the high gain region, and heavy feedback passed through the quartz flats to the input oscillator. We had a proverbial tiger by the tail, and something had to be done.

By replacing the input oscillator with a mirror, the feedback was reflected back through the three-pass XLD-1 amplifier. This so-called "super radiant" system produced 250 kilowatts, the highest output measured to date. The output movies showed an intense beam, suggesting that perhaps the calorimeter was not measuring all of the power produced.

By the second week of March 1969, targets were set up at the target wall at 50 feet. Tests were performed on aluminum parts from a B26 and a titanium nacelle from an SR 71. Holes of various sizes were punched in all targets. Piggyback tests were performed by the Lincoln Laboratory on potassium chloride window material. The targets were located at the focus of the external telescope to obtain the highest possible flux, and both samples were disintegrated. The maximum power achieved in this test was 270 kilowatts.

In early April, the first test with the new cooled optics and the hardened input system produced 330 kilowatts. Feedback damage was reduced significantly. Shadowgraphs showed strong wakes down into the supersonic cavity, which were being formed by the lands that had been used to position the nozzle wafers. The streaks that had been observed previously in the output beam evidently had been caused by the wakes. The streak problem was improved by reducing the number of lands. Eventually, in a few years, they were eliminated entirely by the use of fewer, much thicker wafers that were stiff enough to hold position without any lands. By the end of April, a new maximum power level of 400 kilowatts was achieved.

To fire the laser beam at targets from long distances, a telescope to focus the beam would be required. In Florida, some sort of protection from the weather was required for the telescope, and I remember the telephone conversation with the people at the Weapons Laboratory. The technicians wanted the telescope and agreed that an enclosure was needed. I explained that we had some choices; however, if the enclosure were constructed from stainless steel, it would be expensive. With their tight budget, the technicians did not want that option. A much lower-cost alternative was to construct the enclosure from cinder block. The contracting officer struggled with this but eventually accepted that choice because no capital budget existed for such things. We explained that the structure was to be called a telescope enclosure and never referred to as a building, and all agreed. Shades of the model support in Chapter 1.

The telescope enclosure was completed by early May 1969. Some time later, Alden Smith who now was head of operations, toured the E9 facility with me. He had not been there for quite a while. Among all of the things about which Alden Smith had to worry, capital appropriations and expenditures were among them. As we drove around the end of E9, he said, "What's that building? I don't remember that." We had to tell him that it was a telescope enclosure, not a building. We explained the details, but he was shaken anyway.

With the desire to propagate the beam longer distances through the atmosphere, the question of thermal blooming increased in importance. When the high-powered beam passed through still atmosphere, it tended to heat the air and defocus the beam. To help with important propagation questions and to provide an interface with their world, another scientist from UARL joined us. Ed Pinsley noted in our interview for this book that although our Pratt & Whitney performance analysts, under Chuck Staley, did an outstanding job of calculating laser and optical performance, the eyes of our fearless government scientists would glaze over because a Ph.D. title was not behind Chuck's name. Dr. Edward Sziklas could speak on their terms, which solved that problem. However, one thing Ed Sziklas could not fix was the government scientists' complete misunderstanding of the need for spare parts in the development of complex machinery. They thought that if the physics were correct, we should be able to complete the job with one set of parts. The agonizing part was their equally galling pique when a delay resulted from a part failure.

Ed Sziklas, one of my closest friends, is a world-class physicist and did a lot of original work on thermal blooming. Thermal blooming is a problem that occurs if a laser beam is passed through a particular column of air for any period of time. The heated air tends to act as a lens and distort the passage of the beam. Ed determined the extent of the distortion and suggested ways in which it might be compensated. UARL was under contract with the Office of Naval Research to devise ways to reduce thermal blooming. It was discovered that with feedback from the target controlling a deformable mirror in the output system, it was possible to precondition the beam and significantly reduce the problem.

While we were still interested in increasing the power output of the XLD-1, the optical quality of the beam was increasing in importance. A 20-cps beam wander was solved quickly by stiffening the input system mirror mounts. The streaking problem would take much longer to resolve. The new Barnes camera was used to record the beam profile at high flux levels by observing reflections from a polished copper target. A series of piggyback tests were performed by both the Lincoln Laboratory and the U.S. Air Force. The Lincoln Laboratory at Massachusetts Institute of Technology continued to evaluate potassium chloride crystals as a potential window material. All of the tested samples cracked. The Air Force tested an aluminum airfoil section, a laminated nylon phenolic, and a Plexiglas® windscreen. The windscreen and the airfoil were penetrated in one-third to one-half of a second, and a deep crater was vaporized in the surface of the phenolic. The laser strikes a surface with a heat flux as a radiant beam from a temperature source many times hotter than the surface of the sun. By the end of July, the power level was up to 460 kilowatts, which was to be the record through the end of 1969.

In addition to the distortion caused by the wakes from the separating lands, leakage around the ends of the nozzle caused significant beam spreading, measured to be 20 times the ideal or diffraction limit. End plates had been cast integral with the wafers but did not significantly reduce leakage, and a rubber sealant called RTV was used on the next build. It was effective, provided a good seal, and substantially cleaned up the flow. That was the good news. The bad news was that the sealant also contaminated the mirrors with small bits of sticky material and had to be abandoned.

A night shot was a memorable thing to see in Florida. The radiation in the beam had the characteristic 10.6µ (infrared) wavelength of the carbon dioxide laser and under normal circumstances would be invisible. Two things made the beam a visible bright flash, similar to those seen in the movie "Star Wars." The first thing was the streaks that resulted from the aerodynamic distortion, and the second was from the bright flashes from the incineration of dust particles and insects that inadvertently flew into the beam. This probably was the largest and most expensive "bug zapper" in the world. Figure 7.4 was taken during a diagnostic test, with the beam passing over an optical bench. The fuzzy streaks are produced by the incandescent dust particles.

Figure 7.4 A laser beam at 600 kilowatts. (Courtesy of Pratt & Whitney)

After I had left the program, a range had been constructed with a fancy railroad to translate the targets across the beam, approximately 2 kilometers downstream from the XLD-1 and near the boundary of the Pratt & Whitney property. The main north-south power transmission line for FPL crossed Pratt & Whitney property only 500 feet or so from the laser. The thought struck me that if South Florida were reported to have been plunged into darkness some night, perhaps the boys had gotten their elevation a bit too high.

One night in October after 377 firings, the tie-bars failed and unzipped the XLD-1. One major difficulty encountered throughout the project had been that the scientist in the government who was managing and advising on the program did not have any feel for the problems inevitably encountered in the development of large, highly stressed, hot hardware. This had been an initial rig to demonstrate feasibility; a follow-on water-cooled design for the long term had been planed but never funded. After much hand wringing by the government committee, Dick Coar located the money to rebuild the XLD-1.

In November, an RFP from AFWL started what became the Airborne Laser Laboratory. Pratt & Whitney built the laser that was flown in a revamped KC-135 with a Hughes tracking and optical system to deliver the beam. Ultimately, the aircraft was to shoot down a number of drones and U.S. Navy missiles. The story of this program is told from the viewpoint of the U.S. Air Force in the book titled *Airborne Lasers: Bullets of Light*, written by Robert W. Duffner and published by Plenum Press.

The new XLD-1 was tested for the first time in February 1970 and was a big improvement. In early March, a power of 540 kilowatts was demonstrated, which exceeded the contract requirement. We had published a catalog on the mirrors available from our optical laboratory. This part of the business had developed from the original start previously described in this chapter. The mirrors now were made of molybdenum rather than copper—a much harder surface and easier to polish. They also were gold coated to boost reflectivity. In March, we received a contract from the Lincoln Laboratory for two 20-inch diameter mirrors that used the rocket cooling technology called THERMAL SKIN®, invented in earlier days by Chuck Staley for the high-pressure rocket as a possible alternative to transpiration cooling.

March 16 was the last weekly laser project report I signed. Henceforth, Ed Pinsley took over both the signing and writing of the report. The project continued and grew, and Bill Missimer directed the A11 program to supply a flyable laser to the Air Force. In the fortieth anniversary memorial tape produced by Pratt & Whitney, Bill relates the story of an initially unexplainable delay that occurred during a high-powered test shot at a distant target. The reason for the delay was discovered after the test. The protective door that covered the exit of the telescope had failed to open on time, and the beam had to punch a hole through the door before reaching the target.

Don Witt spearheaded a later program that employed halogen chemistry to avoid the Achilles' heel of the carbon dioxide laser, which was the inability to penetrate the atmosphere on rainy days. Operation at the reduced wavelength provided by this material improved propagation; however, after many years and company dollars invested, United Technologies Corporation decided to depart the laser field entirely, except for work in what was UARL but is now the United Technologies Research Center (UTRC). The E8 stand has continued to be used in the alternate turbopump development work. However, all traces of the E9 facility have now been completely removed. Only a grass field remains there.

CHAPTER 8

The Space Shuttle Engine

In the spring of 1969, NASA asked potential contractors whether engines could be developed for a reusable launch vehicle to fly by 1974. Lowering the immense launch costs by having a reusable space truck seemed to be a saleable target. Landing on the moon was a tough act to follow.

The advanced rocket engine development work at Rocketdyne, the division of Rockwell International that had enjoyed almost all of the Apollo engine work, had been focused on a concept called the Aerospike. While Pratt & Whitney was developing high-pressure staged combustion, Rocketdyne had been pursuing this completely different alternative concept. It had been supported by contracts from both the U.S. Air Force and NASA. The Aerospike can be visualized as a bell nozzle turned inside out. The throat, which would have approximately the same total area as an equivalent bell nozzle engine, is stretched to a long narrow circular slit near the maximum diameter of the engine. The ideal nozzle expansion surface becomes a conical point. In theory, because there is no surface on the outside of the engine exhaust, external air can flow inward toward the plug to compensate for pressure changes, and overexpansion cannot occur. Prior to October 1969, Rocketdyne promoted the Aerospike as its offer for a reusable engine to NASA.

For the past decade, Pratt & Whitney's advanced work had been devoted to the high-pressure staged combustion engine. As outlined in Appendix B ("Space Transport Engines"), we had anticipated that a space truck would be built. Finally, our time seemed to have arrived.

Figure 8.1 represents the development work done in various sizes for each of the major components of such an engine. The numbers within the rectangular bars represent the thrust levels in which the different components were tested (i.e., 250K = 250,000 pounds of thrust). The handwritten numbers associated with each of the bars indicate the pressure levels or rise achieved.

The total investment in this work approached one-half billion of today's dollars and was shared by Pratt & Whitney, the U.S. Air Force, and NASA. This represents the most exhaustive technology development effort ever directed toward a specific engine type, the long-life space transport engine.

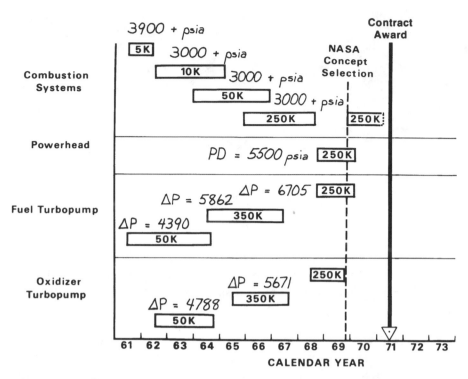

Figure 8.1 Critical component test history. (Courtesy of Pratt & Whitney)

However, Fig. 8.1 does not convey the tremendous difficulty of this work nor the dedication of the people who accomplished it—not only the designers and engineers but also the test stand operators, technicians, and everyone involved. The mechanics in the shop that had made the parts being tested wanted to know how things were going. Some of the best and the brightest talent fought to join the group, realizing we truly were conquering a new frontier. They gave their all. The opportunity to use new approaches to correct mistakes made in previous designs is the secret to rapid learning, particularly in a stable and cohesive group. These same people had worked on a series of the jobs, which provided the essential continuity. The 350K fuel turbopump avoided some of the disasters of the 50K pump; however, it was a long and difficult struggle to meet the design pressure rise, requiring multiple builds and two years of elapsed time. A breakthrough occurred when the 250K turbopump performed exactly as designed with one set of parts in less than six months. It also achieved a record for liquid-hydrogen pressure rise of 6705 psia. As I look back, I think we made it all appear too easy and gave the impression that anyone could do it.

Everyone remembers the exhilaration when we succeeded. However, they also recall the many nights spent in the E8 control room, kneeling on the floor and poring over high-speed strip charts as we struggled to make the system work. The O-graphs that recorded high-speed data produced paper rolls that were 500 feet long. When spread out, these rolls ran across the control room, out the door, and down the hall.

Figure 8.2 is a 1960 photograph of the rocket area and shows the relative position of the control room, the north and south test areas, and the dual engine vertical stand used during the RL10 development. Two new vertical stands were built behind the control room for the RL10. In late 1965, construction began on the E8 high-pressure stand in the south test area. For safety reasons, the distance between the control room building, constructed as a reinforced bunker, is approximately a quarter mile from any of the test stands. Working with rocket propellants and gases at pressures of 10,000 pounds per square inch requires all testing to be done remotely. All parties retired to the control room building after the test preparations on the stand were completed. The only windows in the reinforced walls were located over the test operator's console, and they were heavily constructed and aimed at the particular test complex in use. Before the age of remote television, a surplus submarine periscope was mounted through the roof to obtain a close-up view of the test stands. When a firing such as that shown in Fig. 8.3 was made, the noise level at the engine exceeded 165 decibels. It was a smooth noise with little ripping or tearing at low frequencies, but it could be felt on the chest, as well as heard.

Without realizing it, we accidentally invented flextime. When we had a tough problem in trying to get a firing off, people would work two or three days without going home, sleeping on tables in the lunchroom. After that, they would disappear for a day or two, and I don't remember

Figure 8.2 The rocket test area. (Courtesy of Pratt & Whitney)

Figure 8.3 Firing of the XLR129. (Courtesy of Pratt & Whitney)

their absences ever being discussed. Everyone was focused on getting this new technology to work. The hardware was becoming pretty good by 1969 or 1970, with few show-stopping mechanical problems. However, the sheer complexity of the engine and the test stand—coupled with the high metabolic rate encountered when 5 million horsepower is stuffed into something the size of a wastebasket—required great care.

Del Tischler, who headed the Propulsion Section of the Office of Science and Technology (OAST) kept his hand on the pulse of all advanced rocket work that was happening in the country. One of Del's jobs was to make presentations on the status of this work to the Congressional committees and their staffs responsible for the NASA budget. Del polled each of the contractors to obtain their latest data and viewgraphs. The testimonies are in the Congressional Extension of Remarks.

NASA examined all of the data for the supposed benefits of the Aerospike and found it wanting. In October, NASA told Rocketdyne to switch to the high-pressure staged combustion engine. NASA did not call this a Pratt & Whitney concept, but that is what it meant. One of the Rocketdyne executives said, "We felt as if we were running fourth in a three-man race." I wonder if anyone told them not to worry?

At the Twenty-Fifth Annual Meeting of the International Air Transport Association on October 23, 1969, at Amsterdam, NASA Associate Administrator of Manned Space Flight, Dr. George E. Miller, delivered a historic speech in which he talked about the future of our space program. The speech, placed by Honorable Olin E. Teague into the Congressional Record on page 37782 of the Extension of Remarks of the Congressional Record, December 8, 1969, covered all aspects of the future space program. Part of page 37784 particularly caught my attention because it related to the propulsion system:

> The high pressure, staged combustion rocket engines for the space shuttle will differ from the expendable engines used in our previous launch vehicles since they will be designed to have many of the characteristics of the engines in use in modern jet aircraft, including stability over a wide range of operating conditions, variable thrust to permit vehicle control, time between overhauls measured in hours of operation, high performance and operational dependability.

Comparison of these words of NASA (October 1969) with those in the second paragraph of "Space Transport Engines" (Appendix B) published six years earlier (September 20, 1960) and those of Chapter 5 reveal a remarkable similarity.

> To provide an economic space transportation system, it appears that the next generation of launch vehicles will be developed as reusable systems. Engines for these vehicles will be significantly different from the throw-away types in current use, and must possess many of the characteristics of the engines in use in today's jet aircraft. Stability over a wide range of operating conditions, variable thrust to permit vehicle control and ground checkout, time between overhauls measured in hours, and operational dependability must be coupled with high performance to meet the propulsion needs of a space transport system.

It appears that our message had reached the power center of the rocket engine world. The lack of attribution was noticeable, but at least our concept had been embraced.

This was the time that the XLR129 hydrogen turbopump was being tested with the big heat exchanger driving rig in Area B, as described in Chapter 6. In addition the new preburner and power head case was being checked in preparation for the test with the turbopump also described in Chapter 6. This test series, completed in August 1970, was the last under the U.S. Air Force, and the contract was transferred to NASA.

In September, the group was reshuffled, in name mostly to satisfy the forthcoming formal Space Shuttle Main Engine (SSME) Phase C/D competition. I became the senior program manager for the SSME, and Bob Atherton was the deputy for engineering. Frank McAbee returned to the group as deputy for marketing. The tests of the XLR129 under NASA contract were conducted in February and March 1971. This was a series of fourteen tests, for a total of 251 seconds of a complete staged combustion system and two-position nozzle with an improved main injector and turbopump simulators. The 250K liquid oxygen turbopump that would have been on hand had been cancelled by NASA. The fuel turbopump previously had been tested as a component with the preburner and the main case. Without the liquid oxygen side being available, control complexity, particularly during the starting transient, dictated that both propellant supplies be pressurized.

The I_{sp} performance measured with the new injector improved over the original configuration by 4.5 seconds, or almost 1%. As described in previous chapters, the objective for the transpiration cooling system was to develop a design with low loss coupled with unlimited life. This objective was reinforced by the RFP for the SSME issued by NASA on March 2, 1971. The requirement was that every part must have at least 100 cycles demonstrated by test and an analytically proven design life of at least 400 cycles. NASA truly was committed to a long-life reusable engine. To ensure that NASA received what it asked for, demonstrator engines were to be built by the two finalists and tested, and then the choice of contractor would be made. Exhaustive design studies had convinced us that convectively cooled, milled-channel chambers could not survive the high number of cycles required. The high temperature gradients that occur between slots and the repeated hot-cold cycling inevitably would lead to cracking of the coolant passage long before the required 400 cycles had been achieved. I believe the failure mode may have been called blanching.

We would remember for a long time another part of the RFP that was a condition for entering the competition. This condition was that the contractor must agree to relinquish all associated patent rights to NASA. In retrospect, our naiveté was unbelievable.

Anything that introduces non-uniformity in mixture ratio across a rocket exhaust stream reduces performance. Transpiration cooling with the injection of a modest amount of unmixed hydrogen through the wall unquestionably produces some minor performance reduction. We thought it best to make a direct measurement of this effect. Therefore, we constructed a milled channel chamber that could be installed in place of the transpiration-cooled chamber. The final six tests of the XLR129 program totaled 90 seconds and were made with this chamber. These tests proved that the penalty for transpiration cooling for the Orbiter engine was 0.5 second and 1.2 seconds for the booster engine, which seemed a reasonable tradeoff for unlimited life. At this time, the vehicle configuration envisioned two completely reusable stages, employing the same core engine in appropriate numbers for each stage, with the nozzle expansion ratio tailored for each stage.

Carl Kah, the project engineer on the XLR129, is shown in Fig. 8.4 as the man in the hardhat. Carl is an inventive individual and does not give up on a difficult challenge. It was desired to sample the exhaust gases across the flow at the exit of the nozzle. This type of sampling had been accomplished successfully with air-breathing jet engines. The F100 afterburning fighter engine exhaust stream has a total pressure of slightly less than 3 atmospheres at 3000°F. Now we wanted to capture samples in a gas stream moving at Mach 6, with a total pressure approaching 200 atmospheres and a total temperature of 6000°F. This was a difficult environment, but Carl did it with transpiration cooling of the probe that was interrupted at the moment of capture.

Figure 8.4 The XLR129 on the E8 test stand. (Courtesy of Pratt & Whitney)

Frank McAbee was convinced that our major competitor was very active politically and thus we should try to tell our story to those who were central to this struggle. We had an appointment to meet the chief administrative assistant of Senator John Stennis at the Petroleum Club in Jackson, Mississippi. As a group, the administrative assistants are some of the brightest people I have ever met. The first time Frank had pointed out one to me, the administrative assistant was walking and whispering into the ear of an octogenarian senator shuffling down the hall in the Senate on the way to a vote. Our political system, for good or bad, depends heavily on the skill of these people. The extent of the political coverage of our competitors was apparent to us on these trips. As we walked into the club, our host asked, "Who is your competitor?" We replied, "Rocketdyne." Then the host said, "Well, let's have a nice lunch." That was the end of our discussion concerning the competition. Our effort to tell our story was too late.

I have neglected to mention our third competitor, Aerojet. It is a major rocket engine contractor for the U.S. Air Force and built the storable liquid engines for the Titan, among other things. Aerojet had adopted the high-pressure staged combustion cycle early in the competition and had done some exploratory test work with high-pressure liquid hydrogen. The company had no more chance than we did but was merely window dressing for this faked competition. I remember the Aerojet people as being nice folks.

On a lighter note on a similar mission, Frank and I were driving along a back road in Mississippi in a rental car when the car ran out of gas. We had seen a general store earlier during our ride and decided to walk there to obtain help. I am from southeastern Massachusetts, and some would say you can discern that from my accent. Frank is from Alabama and can speak "good ol' boy" with the best. Frank suggested that I not open my mouth at the store because anything I said would come out as "Yankee." I did as I was instructed and listened to a conversation I barely understood. In short order, we were given a milk can that was half full of gasoline, and we headed back to the car with Frank's promise that he would return the can. Had my origin been detected, we would still be walking. I bet the milk delivered in that can tasted funny for a long time afterward.

Congressman Eddie Hebert from Louisiana was chairman of the House Armed Services Committee, and he was the most fun to visit. He had what was called a "safe seat" and would regale us with stories of Southern politics. Eddie would talk to us about our problems and was very kind, but we always felt that our stories had not moved him much.

During the design phase of the proposal effort, Huntsville asked for additional details regarding how we had run almost every test, supposedly to substantiate our design decisions. Finally, we received a request for a reference library of all tests we had run on every component. We were busy with the proposal work and gave the so-called "reference library" short shrift. We did not need it. We knew how we had run the tests. Attribute this again to our naiveté, but we did not realize the panic our abrupt answers had caused in Huntsville. Those in the know suspected that we might have tumbled onto what was happening. They should not have worried. We knew how difficult the job was and that no one else had the experience to accomplish it in the proposed time. When we realized how serious those at Huntsville were about wanting the data, a complete job was done, and they breathed a sigh of relief.

For the past ten years, we had visited each of the vehicle companies and the cognizant government agencies many times—at least twice a year—to keep them up to date on the performance potential and the experimental success we were having with the technology. It was necessary to convince everybody that there were significant benefits and that it was practical to work at the high-power densities. Unfortunately, with this mantra that high pressure was practical, it had not occurred to us until too late that it had become necessary to convince them that it also was a difficult task to accomplish. The corollary was that Pratt & Whitney was the only group of people in the United States who had done much of it.

In terms of the bottom line, rockets were not a significant part of Pratt & Whitney's business. The top management always wanted somebody to pursue advanced work, and the middle management would tolerate the presence as long as the effort did not impede our real business, which traditionally had been large air-breathing engines. It was difficult to convince our immediate brass to accompany us on our forays to talk to customers. The big brass would go along sometimes, and I can remember one trip to Texas to meet with George Low, a former professor who was NASA program manager for the shuttle at that time. Bill Gwinn, the chairman of United Aircraft Corporation, accompanied me. It was now late in the competition, and the presentation tried to emphasize how difficult some of the technology was to accomplish, such as the almost 7000 psi discharge pressure of the cryogenic pumps. I can remember that Low was sitting behind his desk, with us in front. He became disturbed and turned his swivel chair sideways, saying, "If I believed what you're telling me, we could have made the decision before the competition started." I was chilled by the tone of his voice. As we left, Bill Gwinn said something to me that sounded like "...that didn't sound too good!"

One of the critical points in big competitions is an event called the Critical Design Review (CDR). It is a series of oral presentations to substantiate every discipline in the design being submitted, similar to a dissertation for a doctorate degree. Our leaders for each part of the engine had practiced their presentations, and everyone felt confident. Technically, we knew what we were doing.

Almost coincident with our competition, another competition underway was much more significant to Pratt & Whitney. The F100 engine would be used to power the F15 and later the F16. Bill Gorton held a meeting every morning with the key players on that program. Knowing the importance of that struggle and the tightness of his time, I was surprised when Bill's secretary said he wanted me to go into the meeting before I left. We were gratified that Bill would interrupt the meeting for us and assumed he wanted to wish us well and speed us on our way to the CDR. We had been planning to fire the 250K on the E8 stand during the scheduled visit of the NASA source selection boards in approximately two weeks. I went into the meeting, all pumped up to go to Huntsville. What Bill did was to reprimand me for causing the overhead rate to increase and stop the preparations for the visit. Any increase in overhead was a negative for the F100 competition. The relative status in the company of the jet engine and rockets was sharply etched. It was hard to recapture the former upbeat mood.

As an alternative to firing the engine, the fallback position was to show a movie of the firing. At least the viewing could be tightly scheduled, whereas a firing could not. However, it would not be the same as a demonstration of the 165-decibel real thing.

Looking back, as I have already remarked, our naiveté is difficult to imagine. Having grown up in the sheltered arena of working for the top secret community, where most decisions were based on technical facts, I was unprepared for the world of spin in which your connections or your presentation were more important than what you did. We knew that only Pratt & Whitney had the background to do the job in the time required. We thought that the one-half-billion-dollar effort (in today's dollars) over the decade would, in contrast to no work in this area by the competitors, be the determining factor. The faith was shaken a bit when Dale D. Meyers, Rockwell's vice president and manager of its Space Shuttle program, was named Associate Administrator for Manned Space Flight by NASA. In this position, Dale could hand-pick the source selection board for the SSME and review its findings in secret.

139

NASA also announced that the planned hardware demonstration of the two finalists was cancelled because it did not have funds to support the effort. The selection of the winner now would be solely at the discretion of the administrator.

The visit of the source selection board was one of the most uncomfortable days I have ever experienced. Except for Colonel Walt Moe, from the U.S. Air Force Rocket Laboratory, the assemblage seemed either disinterested or hostile. I guided Eberhardt Reese, who had succeeded Wernher Von Braun as director of the Marshal Space Flight Center, through the display of the 250K turbopump hardware that had achieved the record 6705 psi. The parts sparkled like jewels, and, although Reese said little, his facial expression looked as if he smelled something foul. Willie Maratzek, one member of the source selection board from Marshal, was so vocal about his dislike of transpiration cooling that Herman Widner, who was acting as "den mother" for the group, felt compelled to quiet him. It was unfortunate that we could not fire the XLR129 for them, because that would have grabbed their attention even if it did not change any votes.

The initial Space Shuttle plan called for two completely reusable oxygen/hydrogen stages, with the same engine core used in both stages and with the nozzles tailored for the particular stage. In a later economy drive, solids were substituted for the reusable first stage. Pratt & Whitney's proposal (Fig. 8.5) incorporated a two-position nozzle for the second stage.

This design was the culmination of a ten-year effort and represented the best of what had been learned. The cooling system could meet the required 400 cycles. The spherical case design was an elegant solution to some extremely difficult structural and flow distribution problems and had been demonstrated in the 250K hardware. The radial spray bar configuration of the main injector had demonstrated by test the combination of excellent

Figure 8.5 Mockup of the proposed Space Shuttle Main Engine. (Courtesy of Pratt & Whitney)

performance and rugged durability. The two-position nozzle that is the bill of materials in three different models of the RL10 has proven in flight to be very effective. One of the most important choices, as history would prove, was the configuration of the liquid oxygen turbopump. It was based on the successful single-entry concept of the 350K high-pressure turbopump that had been developed under contract to NASA. Jumping a bit ahead in our story, the ease of that program soon was shown to be in direct contrast to the agonies that would be encountered by Rocketdyne with the double-entry pump. The lesson learned during the work described in Chapter 5 was that a rugged hydrostatic thrust piston design, which guaranteed the axial location of the rotor inside the cases, was an excellent trade for a few inches of length and pounds.

All of the learning accumulated in the ten years of effort was embodied in this engine design. It was a masterpiece. The test hardware routinely was operated as long as the capacity of the test stand tanks would allow, at full rated thrust. That this engine was not to be developed was a great loss to the United States. In my opinion, missing that chance was to cost many extra billions of dollars.

The proposal was submitted to NASA-Marshal on April 21, 1971. The first engine firing was to have occurred in the first quarter of 1973.

Richard Bissell retired from the CIA and was a consultant for United Aircraft. During the interregnum, he paid me a visit. He was a kind man and, after some conversation, said, "I am sorry, but you cannot win. It was already decided in advance. The only reason for the competition was to transfer your technology to them." I wish Richard Bissell had visited me a year earlier. It would have been interesting to watch NASA scramble with its future, having become so totally dependent on the concept.

On July 13, I received a phone call from Senator Claude Kirk's administrative assistant who could barely contain the smile in his voice when he told me we had lost to Rocketdyne by only six points out of a thousand. You can imagine the rage felt by our group in knowing that what should have been ours was being taken away after our ten-year struggle to make it work.

We then received a letter from the NASA administrator at that time, Dr. James Fletcher, explaining that Rocketdyne had done some important work at full thrust. Even in our blinded state, we did not understand why this one short firing could be so important to him. Essentially, it was the same sort of simple exploratory test that we had run ten years earlier. The size was chosen to match the thrust requirements of NASA and was of no technical significance. It was a larger piece of copper but of no real importance.

After the official NASA history writer had sent me some draft notes, everything became clear. One pressurized firing had been made for a duration of 0.46 seconds, long enough to get a picture. The decision was made not to make a second firing because of the risk of erosion of the chamber. When the picture was shown in a presentation to Eberhardt Reese, he stood up and remarked, "Now I really believe it can be done." Both Reese and Fletcher needed cover. If questioned now, each could respond to the uninitiated that Rocketdyne had fired full-size hardware.

CHAPTER 9

A Cry for Help

The loss of the engine competition for the Space Shuttle was a crushing blow for everyone at Pratt & Whitney who had worked so hard for so long to make possible the propulsion for the Space Shuttle. The mood of these people working on the rocket business contrasted sharply the mood of those in the jet engine business at Pratt & Whitney. Those in the jet engine business recently had won the F100/F401 program to develop and build the most advanced jet engines of the day to power the F15 for the U.S. Air Force and the F14 for the U.S. Navy. The projects would mean many more dollars in revenue and were much more important to Pratt & Whitney than was the Space Shuttle project.

The silence from all of the vehicle contractors regarding the injustice of the SSME award was deafening. They knew but dared not bite the hand that fed them. We had thought NASA would realize that the development of such an engine would be difficult. When our recommended path, laid out six years earlier, had been followed in such detail, it seemed as if our turn had arrived. What we did not anticipate was the arrogance of those who had been to the moon. They could do anything and knew no limits. However, this job turned out to be tougher than they expected. To provide the best possible performance obtainable from the hydrogen/oxygen propellants required all components to operate at high power density levels. That is, the metabolic rates were approximately five times greater than anything NASA or Rocketdyne had faced in the past. During the decade in which we had done this work, we had endured many spectacular failures. However, after the half-billion dollars of effort, we finally were getting things right. A large piece of copper and the abbreviated 0.46-second demonstration test described in the preceding chapter did not prepare Rocketdyne for the future.

Several years later, Del Tischler asked why we had never explained to NASA management how hard the job would be. Del understood the difficulty, having followed the work so closely for many years. I think the answer is that after struggling for ten years to convince everyone that it was possible to work at such high metabolic rates, having been down that path ourselves three times in the past, we knew we could do it. The thought of someone else starting essentially from scratch was not considered. The Pratt & Whitney pitch continued in the vein that it was possible to make it work, as our testing had proven. Our failure to sense that the game had changed and that the customer's program plans depended totally on the success of the concept was our loss. We did not recognize that our strategy to win had changed. It probably would not have made any difference anyway. George Low's response to the message of difficulty was that NASA simply did not want to believe the facts. Similar situations had happened previously where a winner with little specific experience was forced because of time constraints to proceed with the development of a cartoon drawn in haste during a proposal and later encountered vast difficulties. NASA and Rocketdyne were in trouble.

Maybe George Low, NASA program manager for the Space Shuttle, did hear the message. Directly after our loss, Karl Heimburg of the Huntsville technical staff approached Gene Lehmann, our local representative at the time, to ask if Pratt & Whitney would accept a contract to develop the turbopumps for Rocketdyne's engine. Dr. Fletcher called Al Horwath, who ran the Washington, DC, office at the time, to make the same offer. Perhaps they did not understand the depth of our feelings at the time and thought that such an offer might be a cheap way to turn our wrath. Maybe some of them were seriously worried about how difficult the future might be. Although Pratt & Whitney many years later would build the turbopumps for the Space Shuttle, the group certainly could not have accepted the job at that time because of feelings of a terrible injustice that had been inflicted. The rage also was felt at the top of United Aircraft Corporation, and it was the president who drove the protest effort, the first in 45 years of doing business with the government.

The United Aircraft protest was orchestrated by the best legal help in Washington, DC, for these matters. The law firm of Reavis, Pogue, Neal, and Rose had an excellent success record. Frank McAbee and I provided the Pratt & Whitney input, and we developed the greatest respect for "Took" Crowell and Stan Johnson, who handled the case. We would fly to Washington every week for ten weeks, stay in the Madison Hotel, and return to Florida on the last flight every Friday night. The initial rage had been eased by time, as well as by Took and Stan's cool logical approach. However, we felt frustration when arguments that seemed straightforward from an engineer's perspective did not fit into a legal framework. NASA had done nothing illegal, particularly when it had required us at the outset of the competition to relinquish all rights to our work. Del Tischler's testimony before the Congressional committees contained in the record for the late 1960s clearly shows who had done what. No distinction had been made between one demonstration firing for 0.46 seconds and a decade of exhaustive development of all of the difficult technologies—at least in the legal arena. Elmer Stats, director of the General Accounting Office, announced on March 31, 1972, that the protest was denied and found to have no merit. Fortunately, time can be a fairer judge.

Inadvertently, an unanticipated effect occurred. The protest had built a fence around our background data, although we had been forced to grant free access. The normal human syndrome "Not Invented Here" would have protected most of it, but no one could now admit that it was useful at all. Stan had pointed out the flaws in their approach and had detailed the Pratt & Whitney route. History would prove there was a difference.

How did it happen? This is a question with which we have struggled for a long time. A conspiracy to take what was ours and give it to someone else evidently had been the plan from the outset. The Machiavellian twist to the detailed grilling Pratt & Whitney had undergone, in which NASA sought to learn every detail we knew about the process, makes us shudder in retrospect. Our group had "bent over backward" to tell everything. It also became apparent that although the top brass of NASA in Washington directed the plot from the outset, the Huntsville organization expected that Rocketdyne would be its engine contractor simply because Rocketdyne always had been. Why was that so?

Del Tischler, the head of NASA advanced propulsion technology from the beginning and a good friend, told me the following story much later. I had asked Del what had developed the close relationship between Rocketdyne and Huntsville. In the early days shortly after World War II, North American Aviation had a contract from the U.S. Air Force to develop a ramjet missile called the Navaho, which required a liquid rocket booster engine. After being turned down by all of the engine companies that were too busy with the jet age, the Air Force gave the job to develop the rocket engine to the Rocketdyne division of North American. Sam Hoffman, who ran that company at the time, decided to approach Wernher Von Braun and the group from Peenemunde who had been brought to the White Sands military base. Who else knew more about liquid rockets? The Germans felt as if they were little more than prisoners of war, and when Sam invited them and their families to California for an extended visit at his expense, their gratitude knew no bounds. Rocketdyne became their engine company.

That may have been part of it, but the relationship had been cemented at a much higher level. Following the Apollo-1 fire on the launch pad at Cape Canaveral, many questions were being asked by the press about what seemed to be unnatural ties between NASA and North American Aviation which included Rocketdyne. Why did one company enjoy approximately 85% of the Apollo funding? In the book published by Doubleday titled *Journey to Tranquility*, the relationship between U.S. Vice President Lyndon B. Johnson, Senator Robert Kerr, and Bobby Black, lobbyist for North American, is detailed. Why did North American build its Tulsa division in the senator's state shortly before the start of the program to the moon?

These things were important, but I have concluded that they were not the only reasons. Who would you choose? The choice here was not something decided after the proposals were submitted; rather, it was decided at the outset. Would you choose the company that had taken you to the moon and would be almost out of business if it lost this work, and whose division president visited you at every opportunity? Or would you choose a company that had the money to develop the technology but was really in the jet engine business? Rocketdyne had covered all political bases and had done everything correctly, including placement of its vice president in the upper reaches of NASA headquarters. Pratt & Whitney truly was not in the rocket business at that time.

The assets devoted to development of the high-pressure staged combustion engine—the manpower and the money—always were held hostage to the air-breathing side if a glitch surfaced there. The investments in rockets had been made as a hedge against the future by upper management of Pratt & Whitney and United Aircraft Corporation. However, when push came to shove, the vital interest simply was not there. Whenever the question arose of a possible separation of the assets, including people, for any new venture into a stand-alone organization to do any new business, the jet engine managers always voted to keep the effort as an appendage of their group. In his interview taped for the fortieth anniversary of FRDC, Dick Coar reflected that if we had won the Space Shuttle engine development work, it would have been in competition with the F100 and we could not have done both. His integrity must have protected the rocketeers on many attempted raids in the past. The old Pratt & Whitney management would never let the liquid rockets develop into a stand-alone business.

Recently, a later generation of United Technologies Corporation (UTC) management has recognized the conflict. The liquid rocket business now is separate from the air-breathing business, although they remain co-located and share the same facility. In addition, company funding has been provided to develop a new higher-thrust RL10. In addition to the RL10 and its derivatives and the Space Shuttle engine work, joint ventures have been forged with the Russian company Ergomesh to import its excellent engines to the United States. On May 24, 2000, the first launch of the new Atlas III was eminently successful. This was the first mission in this country that used the Russian RD-180 engine from Ergomesh, Pratt & Whitney's partner in the joint venture. The main engine in the new Centaur stage in the Atlas III is a single RL10. The division also includes solid rockets for the U.S. Air Force and booster recovery for NASA. The age of space may have arrived at United Technologies Corporation. Unfortunately, it arrived a little too late for this competition.

In the winter of 1971, I returned to East Hartford to a new job as manager of new business development in the marketing department. After 25 years, this was the first time I had been out of engineering, and the adjustment was difficult for me. Dick Coar had preceded me earlier in the year and was the vice president of engineering for the commercial division. Later, Dick was to become executive vice president of the Pratt & Whitney group, and in 1980 I became vice president for technology for the group and again reported to Dick.

After the contract award to Rocketdyne, budgetary constraints forced NASA to change the Space Shuttle concept of two completely reusable stages using the same hydrogen engine core in both stages. Adopted instead was a stage-and-one-half configuration, with a throw-away liquid propellants tank and strap-on solid boosters. So much for ground checkout of the engines before launch. Now astronauts were to be put on firecrackers. After the fuse was lit, there would be no way to shut off the engine. Perhaps in anticipation, George Mueller dropped pre-launch engine checkout from his presentation in Amsterdam, which had been included in our list of desired characteristics.

The early years of the engine development program at Rocketdyne have been chronicled in the Boeing/Rocketdyne house organ called *Propulsion and Power*. The article, titled "27,000 Seconds in Hell," was written by the official NASA historian for the Space Shuttle main engine. The first engine was assembled in 1975, and an ignition test was made in June. In the first eight months, it required thirteen turbopump replacements to eventually run the engine at low thrust for five seconds. The first major problem was a destructive vibration in the hydrogen turbopump, which surfaced in March 1976. NASA invited everyone who might help, including Pratt & Whitney, to attend a large open meeting where the problems were discussed. Professor Ezra Childs from Texas A & M University probably was the most helpful. The H-shaped main case with two preburners always had appeared ungainly to us, with its tortuous gas flow paths. However, the real problem was the long, floppy, three-stage hydrogen pump rotor that was an open invitation to whirl. Two interstage labyrinth seals were located between the three shrouded impellers, with inadequate anti-friction bearings at each end of the stack. With Child's help, these seals were modified to act as hydrostatic bearings and supplied a serendipitous solution by providing the dynamic damping needed by the floppy rotor. My name is mentioned as having attended the meeting and not raising any objections, but I cannot remember having been afforded any opportunity there to raise objections. It was a typical, well-organized NASA meeting.

Around the time that engine tests were scheduled to start at the NASA Stennis facility in Mississippi, component testing was to begin at the company-owned facilities in the Santa Susana Mountains. The liquid oxygen side was first, and very early a fire was caused by a failure in the high-pressure turbopump, which destroyed the test rig and the stand beyond repair. This occurred in late 1976. In March 1977, the fuel pump side experienced a similar failure, and the stand again was destroyed. The facility was shut down. In his testimony to the Senate subcommittee, Dr. Robert Frosch, who was the NASA administrator at the time, explained, "We have found that the best and truest test bed for all major components, and especially turbopumps, is the engine itself." That is particularly true if all of the component stands have been destroyed.

Successive major failures continued to occur. The depth of concern at Marshal Space Flight Center was demonstrated by a phone call from Dr. William R. Lucas to Harry Gray, chairman and CEO of United Technologies, in October 1977. It is a measure of how desperate things had become and of the personal courage of Bill Lucas, the successor to Eberhardt Reese and Wernher Von Braun, who as director could make this call. Bill began by saying, "You have no reason to help after what we did to you, but we need your help." Bill's specific concern was the high-pressure liquid oxygen turbopump because of repeated blowups. He requested that a team at Pratt & Whitney be set up to review the Rocketdyne design—not only for the current problems, but to "…tell me what makes your eyeballs hurt. I'm not asking for a rigorous analysis but a review based on your people's experience." Dick Coar told me to assemble a team and do what was asked. I traveled to Florida where the experts were located, and a group of people, each with many years of experience with high-pressure liquid oxygen turbopumps, was assembled. The team included:

- Tony Akin, rotor dynamic analysis
- Bill Creslein, project engineer for high-pressure turbopumps
- Walt Ledwith, assistant chief design engineer
- Howard McLean, high-pressure engine design, project engineer
- Jim Sandy, designer of liquid oxygen turbopumps
- Bill Poole, bearing analyst
- Dick Mulready

In approximately two weeks, the liquid oxygen turbopump review was completed. Each flaw as we saw it was highlighted, and an objective report was written, dated November 3, 1977. Dick thought that the barbs in the draft of my transmittal letter, although understandable, were a little too sharp. Properly toned down, my letter (Appendix C) and the report were sent to Dr. Lucas in Huntsville on November 7. On November 8, part of the team traveled to Huntsville to present our findings to management. The audience included:

- Dr. Bill Lucas, director, Marshal Space Flight Center

- Dick Smith, deputy director, Marshal Space Flight Center

- Jim Kingsbury, director of science and engineering, Marshal Space Flight Center

- Lee Belew, deputy director of science and engineering, Marshal Space Flight Center

- Bob Lindstrom, program manager, Marshal Space Flight Center Space Shuttle Projects

- Jerry Thomson, chief engineer, Space Shuttle Main Engine

- Bob (J.R.) Thompson, project manager, Space Shuttle Main Engine

Bill Lucas expressed his great appreciation for the efforts of Pratt & Whitney on their behalf, and he acknowledged that they knew they were in serious trouble before they had asked for our help. At the end of the presentation, Bill summarized by saying, "What you have told us is that we have a good centerline."

Dr. Lucas also had sent copies of the report to the NASA propulsion staff in the Washington, DC, headquarters. When the NASA powers at the top learned of this potentially embarrassing incident, they ordered all copies of the report to be collected immediately. Evidently, NASA was unaware that Bill Lucas was going to approach us, an indication of how desperate the engineers had become. It had taken great courage and the willingness to "eat crow" to ask Pratt & Whitney for help. Bill's letter of thanks and appreciation to the team is also included in Appendix C.

The ice had been broken. In mid-December, Jerry Thomson, chief engineer at Huntsville for the Space Shuttle engine, called to request Pratt & Whitney's help with the hydrogen turbopump turbine, two of which had exploded and destroyed engines on recent occasions. Three days later, NASA Lewis called Dick Coar to request our participation on a team to investigate the fuel turbine. We countered with an offer to perform an independent review as we had done on the liquid oxygen turbopump because General Electric was already on the Lewis panel and we chose not to serve with them. This time, our review group was:

- Tony Akin, rotor dynamics analysis
- Bob Atherton, director of rocket programs
- Al Hauser, manager of materials development, East Hartford
- Walt Ledwith, assistant chief design engineer

- Howard McLean, high-pressure rocket engine design project engineer
- Jim Morris, metallurgy
- Bob Sellers, turbine aerodynamic design
- Corinne Stevens, metallurgy
- Dick Mulready

On January 26, our findings were sent to Dr. Lucas (Appendix C). Independently, the NASA Lewis team had come to almost identical conclusions and recommendations. A measure of the severity of the problems in the Space Shuttle engine was that it had to operate almost 500°F above the design turbine inlet temperature and, even at this extremely stressful condition, had achieved only 90% of the thrust rating. Of the 14,000 seconds of test time reported, less than 1,000 were at this modest thrust level. Men were scheduled to fly on this engine in less than two years.

There is an exclusive organization of airline, aircraft, and aerospace industry leaders called the Conquistadors, which meets annually. The meetings usually are held at a nice "watering hole" and are primarily social. However, when these top-level folks get together, there inevitably is some serious conversation. At one meeting, Harry Gray, who as chairman of United Technologies Corporation (including the divisions of Pratt & Whitney and Sikorsky Aircraft among others) was approached by Robert Anderson, then chairman of Rockwell International, of which Rocketdyne was a part. Anderson suggested that the two corporations should consider swapping divisions—Rocketdyne for Sikorsky. His reasoning was that it would be a better fit because United Technologies had Pratt & Whitney in the engine business and thus Rocketdyne would fit into the organization. Plus, helicopters needed gearboxes, and that was a major part of Rockwell's business. Harry was aware of the depth of problems at Canoga Park and declined the kind offer. To "pull my leg," however, Dick Coar told me about the first part but did not mention the rejection, and then he asked what I thought about such a deal. I shuddered. Enough of the disaster scenario was leaking out that we knew about the troubles. The question was: Did Mr. Anderson check his suggestion in the halls of NASA? One thing a government agency never likes to do is admit that it might have been wrong about anything.

It became easier for the engineers in NASA to ask for help and for us to respond. We wanted to help the country and did not want our concept to become distorted and cause a disaster. Jerry Thomson, who was chief engineer at Marshall for the SSME development, called me one day to request a meeting to discuss problems they were experiencing with cracks in turbine blades. He and some of the Marshall engineers met in my office and described the problems they were encountering.

It is the nature of high-pressure staged combustion engines that heating rates are very high and change rapidly during transients. The temperature at steady state of the working fluid—in this case, steam and hydrogen gas—was not very high compared to the turbine inlet temperature of jet engines. The problem was that hot hydrogen gas is the most effective heat transfer medium in the world. On start, the turbine blade surface is hit with hot hydrogen gas, while the fibers in the center of the blade are cold at prestart temperature. The outside fibers tend to grow rapidly in length, while the inside fibers do not yet know it is hot. Huntsville had calculated that the start was causing thermal gradients in the material as high as 40,000°F per inch of thickness. On shutdown, the problem was reversed. While the outside of the blade was bathed in the hydrogen blowing down from the system at near-liquid temperature (–420°F), the internal fibers in the center of the blade were still hot. This gradient was 30,000°F per inch in the opposite direction. This whipsawing was cracking blades in a relatively low number of cycles, and we were asked if we had any material that would resolve the problem. Our latest single crystal would improve the situation by a factor of two. Pratt & Whitney also recommended that the ceramic coating being applied to the blades, supposedly to help the problem, be eliminated. It was an excellent source for crack initiation. They limped along— changing blades when the blades cracked—until 1985, when Pratt & Whitney demonstrated a design (Fig. 9.1) that improved the life by a factor of 30. The answer was straightforward: eliminate the offending thickness by making the blades hollow.

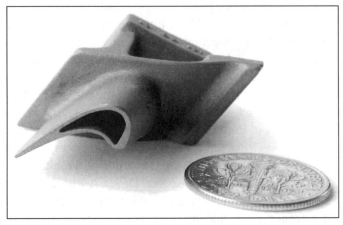

Figure 9.1   Thin-wall turbine blade.

NASA tested the concept in its special thermal cycle test rig that simulated the SSME transient and demonstrated that the Pratt & Whitney design would meet the 55-cycle life goal. This concept was employed in the alternate turbopump development program covered later in this chapter.

In mid-December, Senators Harrison Schmidt and Ted Stevenson of the Senate Science and Technology Committee requested that Dr. Robert Frosch, the NASA administrator at the time, set up an independent review of the Space Shuttle engine by an *ad hoc* committee of the Aeronautics and Space Engineering Board (ASEB) of the National Research Council's Assembly of Engineering. I was a special advisor to this committee, chaired by Professor Gene Covert of Massachusetts Institute of Technology. The reply to the senators was drafted on February 23. The theme was that serious problems existed, and that more time and money was needed. The first manned orbital flight scheduled for June 1979 should be postponed until certain milestones were completed and initial engines should operate derated. The liquid oxygen turbopump should be completely redesigned for later full-thrust engines. When the presentation was made of the covert committee's report to the ASEB, of which Dick Coar was a member, prior to its delivery to the senators, Huntsville minimized the liquid oxygen turbopump problems. Evidently, Bill Lucas had received the word from a higher authority.

Seven more catastrophic failures of the engine occurred by mid-1980.

The engine group at Huntsville and Rocketdyne was learning to live with derating and cracks in parts. The group reasoned that major components would not survive two flight cycles but should survive one. The syndrome became known as the "Black Balls" by some because the liquid oxygen pump ball bearings were turning black after one mission. It was obvious that after one cycle of operation, the liquid oxygen turbopump had to be overhauled. This was the mood when the first flight was made on April 12, 1981. The last major engine explosion had occurred only eight months prior to that time.

Through the largesse of Congress, the coffers opened and money poured on the Space Shuttle project. The engine component stands were rebuilt, and spare parts for the engine were ordered in large quantities to support the engine that needed replacements much more frequently than anticipated—sometimes after every flight. The dream of reusability was put on hold to concentrate on schedule. Through legendary rigorous inspections by NASA before every flight and great care to replace any suspicious parts, the launch program continued, albeit at a much slower rate than had been planned. Although it cost payload, the thrust of the engine was reduced to minimize stresses, and the up-rated levels that had been planned to provide for emergency operation on two engines were dropped. Every flight was a nail-biting experience for everyone concerned about the engine. J.R. Thompson, program manager at Huntsville for the engine, describes in "27,000 Seconds in Hell" their change in philosophy about cracks in critical parts. Having found them in turbine blades and bearings, the trick was to prove that a mission could be completed before they failed. It was a tense time.

The name J.R. Thompson "rang a bell" for me. He later became director of the Huntsville center and retired in 1991 as the deputy administrator of NASA, a winner from the start. I remembered his name as that of one of my test engineers in the RL10 group in the 1960s, and he was one of the first to take advantage of the masters degree program offered by Pratt & Whitney. We thought Bill Gorton, the general manager, would explode when he heard that J.R. had resigned to work for NASA the day after he earned his masters degree.

After both Ed Sziklas and I had retired, we continued to cruise our sailboats together—this time in the waters off the coast of Maine, instead of those of the Bahamas. We were both consulting for United Technologies at the time, and Ed was working on a top-secret program for the Air Force which required special access. From Ed's background, I suspected the work had something to do with lasers. The "den mother" for all technical and scientific consultants to United Technologies was retired general Bill Evans. I had a current top-secret clearance and asked Bill if my name could be placed on the special-access, need-to-know list for Ed's program. Each time I asked, I received a strange answer about why it was impossible; therefore, I gave up. Ed would spend inordinately long periods in the phone booth when we were ashore, but I never questioned him, having been in similar circumstances myself and recognizing the security clamp. One day, Ed did say, "You're going to be surprised when you find out who I'm working with." However, that was all he said.

My wife Carol and I live in Florida in the winter, and she belongs to an antique club there. At one of the club meetings, Carol was talking with Sandy Monk, president of the club. When Sandy recognized Carol's last name, she said, "My husband used to work for yours at Pratt & Whitney!" They had a friendly exchange, and Sandy continued by saying, "We are so happy that Frank will go back to Pratt & Whitney and that we don't have to move to Canoga Park." When Carol came home that evening, she told me about the conversation. Bingo! Sandy's comment had not meant anything special to Carol, but it did to me. The Optical Laboratory, now called OTIL, had grown tremendously since the old days. It was a leader in adaptive optics, among other things, and was the successor to the mirror lab at Pratt & Whitney which by that time had departed the laser business. An unknown buyer recently had acquired OTIL. I now knew who the buyer was, with whom Ed had worked, and why I had been unable to obtain access to Ed's program. If there was one name that was not wanted on the list, it was mine.

Rather than telephoning Ed to confirm all this, I waited until the following spring when Ed and I were sailing again to see his reaction when I said to him, "Ed, I'm surprised that you would work with Rocketdyne!" Ed's mouth fell open, and he said, "How did you find out?" I said, "Carol told me." It had been worth the wait to see his face.

In late 1979, Frank McAbee became division president of the Government Products Division (GPD), the successor to FRDC. Rumors had persisted that NASA had become disillusioned with the ability of Rocketdyne to solve the problems with the engine. During a social occasion with Pratt & Whitney management in East Hartford, a NASA official had mused that NASA wished Pratt & Whitney were still in the rocket engine business. Bearing the scars of the 1971 treatment at their hands, Frank heard this as a cry for help and decided he would try to reclaim some of what was rightfully ours. He assigned Bob Atherton and Jerry Cuffe to study the best approach to NASA.

In late 1982, having always wanted to do some long-distance sailing, I accepted a generous early-retirement package offered by Pratt & Whitney. I reasoned that if I waited until age 65 to retire, the sailing probably would never happen. On April 1, 1983, I was a free agent, and the *Leading Edge* was launched in July. Over the next six years, the vessel covered 18,000 miles between Canada and the Bahamas, and my dream was fulfilled. During the winters of 1983/1984 and 1984/1985, Carol and I lived aboard at a slip in West Palm Beach. With six feet, six inches of headroom, a furnace, two heads, a double bed, and a shower with instant hot water, our accommodations were comfortable and almost complete. However, they lacked laundry facilities, and one day Carol explained to me that the washing machines at the head of the dock were not an adequate substitute. For subsequent winters, we moved ashore to a condominium.

The efforts of Atherton and Cuffe led to an unsolicited proposal to NASA to provide an alternative power head for the SSME. The power head proposal reflected past experience at Pratt & Whitney: that the solution of serious turbine problems often required changes in the upstream systems which provided the hot driving gases, particularly in such a closely coupled system. Such a modification to the engine could be accomplished with relatively minor mechanical complication, with all connections being made at major Rocketdyne flanges. In this circumstance, the internal hydrodynamic designs of each contractor could be maintained. Dick Coar and Harry Gray were behind the effort, and United Technologies Corporation offered to provide the upgrade to the E8 stand to accomplish the job. I jumped at the chance to return to Pratt & Whitney as a consultant. My response was that of the old firehorse when the bell rings.

The sad thing was that Bob Atherton, who had struggled to maintain a rocket presence at Pratt & Whitney during the decade since the loss of the original competition, was stricken with a rapidly moving cancer and did not live to see the rebirth of the business. The company had treated him and his small remaining group of rocketeers as outcasts until Frank breathed new life into the business. Frank hand-delivered the proposal to General Abramson, then an associate administrator, on November 3, 1983. Although the general appeared happy to see Pratt & Whitney in the game again, the general indicated that politics probably would prevent the proposal from being accepted.

In early 1984, Frank returned to UTC corporate headquarters to work directly for Harry. Joe Philips became president of the Government Products Division (GPD), the successor to FRDC. However, Joe's health was bad, and he retired soon thereafter and was followed as president by Jim O'Connor. John Balaguer, the executive vice president, assumed responsibility for the liquid rocket business, and John later succeeded Jim as president of GPD. I hope you have not had difficulty in following the "musical chairs" here.

As predicted by General Abramson, the power-head proposal received a cool response at NASA. It simply could not accept such a major change in the role of its selected contractor. One Friday evening, Jerry Cuffe and I met with Glenn Lunney, associate administrator at NASA headquarters. He explained the political climate and pleaded with us to propose only the turbopumps. NASA had poured approximately a half-billion dollars per year into the problem for a decade, and Rocketdyne had not been able to solve it. Thus, we again were being asked to build pumps for someone else's engine. *Déjà vu.*

Meanwhile, Rocketdyne continued to experience cracking problems with the turbine blades of the liquid oxygen turbopumps, and we had heard some horror stories. We suspected the problem was so severe that it could cause a failure in flight, and Pratt & Whitney decided to offer assistance. Pratt & Whitney's concern was with the political climate. If Rocketdyne suffered a significant in-flight failure before we could provide dependable hardware, national support for the Space Shuttle could be lost. The continuing failures in Rocketdyne's liquid oxygen turbopump represented a real risk. If we did not help Rocketdyne with its current problems, any future opportunities in the business could vanish. I had developed a relationship with Bill Lucas from our prior efforts and was chosen to approach him. Bill was pleased by the gesture and arranged a meeting at Rocketdyne for us to discuss the problem. Our contact at Rocketdyne was R.A. Boudreaux, vice president of engineering and testing.

Attending the meeting on May 7 from our side were Jerry Cuffe, systems; Hal Gibson, project engineering; Dave Lewis, structural analysis, and I. Otto Goetz, one of the brightest NASA people I had ever met, was the Huntsville representative. Joe Strangland, in charge of turbopumps at Rocketdyne, introduced his group. After they became convinced that our motives were pure and that we really wanted to help, they opened up to us.

One particular memory stands out clearly in my mind. As designed, the blade shank/platform intersection area was highly stressed, with almost no margin for vibratory stress. When it was stated that the blade-to-blade platform variations were as standard practice made to fit with a file, an almost audible gasp came from our side of the table. I glanced at Otto, who did not seem to register surprise at the filing. Such crude techniques never would be acceptable in a highly stressed aircraft engine part.

We recognized that these people were from a culture other than that of the aircraft engine business in which engines are expected to operate dependably every day. Instead, in the throwaway world, a single successful operation, each time being the first, was all that was required. Things did not bode well for a long-life reuseable system. To its great credit, Rocketdyne accomplished successive Space Shuttle flights with immature hardware that depended on rigorous inspection and frequent replacement of parts to survive through the short life before infant mortality. Luck must have played a large part. The loss of a single turbine blade during flight probably would have led to a catastrophic explosion. The airlines would never survive with such hardware.

As a result of this meeting, Pratt & Whitney contracted with Rocketdyne to assist with the liquid oxygen pump turbine blade improvement effort. This included recommending design changes and additional testing and instrumentation procedures. Intercession by Pratt & Whitney on behalf of Rocketdyne with the Howmet Corporation, the major producer of turbine blades for Pratt & Whitney which also produced blades for Rocketdyne, also must have been a significant help. We convinced both parties to upgrade the quality requirements of all future SSME blade castings to state-of-the-art standards being used on jet engines. Howmet also was encouraged to pay more attention to the orders from Rocketdyne where the quantities were miniscule compared to those for the jet engine business. Rocketdyne's minimal purchases were an anathma to a busy manufacturer, and it is difficult to obtain a supplier's attention in such circumstances.

The forthcoming RFP requested a proposal for alternative turbopump development, and Rocketdyne was specifically excluded from the competition. In effect, it restricted the changes to be made only to the mechanical design and required the use of the Rocketdyne hydrodynamic design. All flanges were to fit the Rocketdyne engine directly. The turbopumps also must operate with the thermodynamic environment of the engine, which was severe and not realistically quantified. What that meant was that Pratt & Whitney would be forced to employ a double-entry liquid oxygen pump impeller and the long three-stage hydrogen pump rotor, neither of which would have been our choice. Nonetheless, Pratt & Whitney proceeded with a response. What made the difference between the request from NASA now and the invitation made 14 years earlier was that our passions had cooled and the bitter disapointment had subsided. However, we also recognized that NASA had suffered during the decade of terror and it now understood that it needed the help of Pratt & Whitney.

My assignment was to write the executive summary volume, and the work was great fun for me. The format was as follows: The top half of the page would explain the benefits that our design would provide, and the bottom half of the page would define the shortcomings of the current design. Executive summary writing became my forte as a consultant.

A "breadboard" expander-type engine that could be used to test components for an eventual replacement for the RL10 was the subject of an RFP from NASA Lewis. Many smaller programs by other contractors had investigated the possible bits and pieces for such an engine, and it was desired to run them in a flexible, generic engine that could accept components from various sources, such as high-speed pumps. The betting in advance by the *cognizenti* was that Pratt & Whitney had absolutely no chance at winning this job. During the proposal period, our representative in Cleveland was advised by a knowledgeable filling-station attendant to save our time.

Pratt & Whitncy was positive in its response, embracing the approach that would evaluate the benefits of actual hadware and that Pratt & Whitney would help in any way it could. The proposal also noted that Pratt & Whitney was the only organization in the world that produced an operational expander engine. By that time, the RL10 had accumulated thousands of firings and millions of seconds of operation. This experience would be highly valuable in trying to make the breadboard work. Against all odds, Pratt & Whitney won the contract for the breadboard.

Dick Coar used to emphasize the necessity to avoid confusing features with benefits. Upon further examination, NASA concluded that no significant benefits over the RL10 existed, and thus the breadboard program was dropped. This was the latest attempt to supplant the RL10 that has now become a commercial product.

The goal of the Alternate Turbopump Development (ATD) program was to provide turbopumps with a life of 7.5 hours and 55 flights. The contract was awarded in August 1986, and the first firing of the oxidizer turbopump occurred in May 1990.

At the outset of the ATD program, it was clear to Hal Gibson, the engineering manager, that the design of Rocketdyne turbopump housings needed a complete change in philosphy. The housings at that time were pieced together from approximately 200 parts, some castings, and some sheet metal, by hand welding. As an alternative, a new process invented by the Howmet Corporation offered the opportunity to greatly simplify and improve the quality of these major structures. This process, called Microcast-X®, exhibited the fine grain and superior properties of forgings and retained the ability of the casting to produce intricate detail. With an intensive joint project between Pratt & Whitney and Howmet, this type of casting was developed successfully and introduced in the ATD turbopumps. The number of pieces in the cryogenic turbopump housings which had to be welded together was reduced from 200 to only 6. The myriad hand welds required in the old designs were reduced to 4, produced by an automated electron beam system. These changes yielded significant improvement in simplicity and reliability.

The ATD development work was taking longer than had originally been anticipated, and NASA was becoming restive. Grumblings from Congress, possibly inspired by some nonfriendly forces, were heard, criticizing the time and money expended. Dan Golden, the administrator at NASA, set a deadline for completion.

A quirk in the design—a legacy resulting from the inherited double-entry pump configuration—caused the front bearing to be located in a region of low-pressure gaseous oxygen. In this state, the gas available as a bearing coolant has poor heat transfer capability and could not accept even the modest heat generated by the metallic anti-friction bearings. On the watch of Pete Mitchell, who was then engineering program manager, a major technical breakthrough was introduced. The bearings were changed to silicon nitride balls with a diameter of almost one inch. This change from the Pratt & Whitney tradition of metallic bearings was not easy to sell in-house. To some, the idea of using glass balls (their words in an anti-friction bearing), was an anathma. Bob Swinghammer, then the MSFC associate technical director and originally from its Materials Laboratory, initially had proposed the introduction of the nitride. With his strong support, the reactionaries were overcome and the material has proven to be a resounding success. The coefficient of friction of this bearing is much lower than that of the previously used metallic alloys. Together with the 440C races, it has demonstrated extremely low internal heat generation. Cooling with oxygen gas at the state available at this location is now possible. The low heat generation made the difference.

Figure 9.2 Alternate liquid oxygen turbopump for the Space Shuttle. (Courtesy of Pratt & Whitney)

It is interesting to conjecture that this shortfall of the double-entry design may have exacerbated the "Black Balls" problem that plagued Rocketdyne for so long. Pratt & Whitney had proposed a single-entry pump based on the successful 350K liquid oxygen turbopump (Fig. 9.2) developed for NASA which had been troublefree (Chapter 5). Perhaps that would have been an easier path to follow. Silicon nitride also has been introduced in the fuel turbopump for both ball and roller bearings with equal success.

Dan Golden was ecstatic about the fact that Pratt & Whitney had finally solved the problem. At the U.S. Space Foundation Symposium that year, he sought the Pratt & Whitney table and said, "I cannot begin to tell you how pleased I am with Pratt & Whitney. You really turned that pump program around!" The certification occurred in March 1995, and the first Space Shuttle mission with a Pratt & Whitney liquid oxygen turbopump occurred in July of the same year.

The first mission with three Pratt & Whitney liquid oxygen turbopumps was flown in May 1996. Starting in December 1999 and all subsequent missions, the Space Shuttle will have all Pratt & Whitney liquid oxygen turbopumps. These pumps have met or exceeded requirements for a life of 7.5 hours and 55 flights. Operation for full mission duration at 111% thrust, the original emergency rating, also was demonstrated.

The first firing of the fuel turbopump (Fig. 9.3) occurred in February 1990. In August 1991, operation was demonstrated at the 111% power level.

Figure 9.3   The hydrogen turbopump at 111%.   (Courtesy of Pratt & Whitney)

The development work on the hydrogen turbopump was suspended to concentrate on the liquid oxygen side problems. Following certification of the liquid oxygen turbopump, the fuel turbopump (Fig. 9.4) development resumed. John Price is the current program manager for the alternative turbopumps. The Design Certification Review of the fuel turbopump has been completed, and the first Space Shuttle flight with this turbopump is scheduled for February 1, 2001.

Figure 9.4 Alternate hydrogen turbopump for the Space Shuttle.
(Courtesy of Pratt & Whitney)

I owe a tremendous debt of gratitude to the people who worked on these projects at Pratt & Whitney. They were dedicated to finding answers at extraordinary technical frontiers, and they thrived in the thrill of the hunt. Nothing was asked of them that they did not bring through to fruition, usually by breaking new ground. Countless hours were volunteered until solutions were found. From the designers to those in the shop and test stands, this team could not be stopped by technical challenge. The team members can be proud because history has proven their work and the work of their successors to be essential to the country.

L.S. Hobbs's Letter
of December 6, 1939

Luke Hobbs

December 2, 1939

Messrs. A. Willgoos, Insley, Parkins, Cronstedt and Sanbora

It has been fairly obvious from the time of our institution of the Project Engineer system that in reality the system has not functioned as such. That is, the Project Engineers were in effect Development Engineers and did not have a sufficiently broad outlook nor assume enough responsibility. I have been attempting in the various organization changes to gradually raise the standard of our project engineering and the more recent shifts have all been made and approved with the same object still in view. As we are approaching more closely to the desired functioning, I think it is time probably that this be thoroughly understood by all of those concerned and in particular, that the Project Engineers be advised concerning what is expected of them and how they are supposed to function.

It is my idea that insofar as it is possible in a sub-divided large organization, Project Engineers should in effect be Chief Engineers for their particular project. This assumes that they should then have at all times a general knowledge of the entire company situation concerning their project and that their thinking will be guided by this picture in addition to the specific instructions which they are given.

The best way to get at this is to consider the main sub-divisions of activity of a technical business of this sort.

1. Product (Engineering)
2. Sales
3. Manufacturing
4. Quality
5. Service

The Project Engineer should appreciate the functioning of each of these divisions, but what is more to the present point, his particular relation to, and responsibility in connection with, them (I have intentionally omitted finance, bookkeeping and cost accounting, as his overlapping with these is relatively small).

The first item of the "Product", which in essence is "engineering", consists of design, experimental manufacture and test. He is, of course, more intimately tied in here than any other place. Theoretically he should carry responsibility either directly or indirectly for the initial design, should then see that his project is manufactured experimentally and then properly tested. With the exception of design, these are the duties he has been fulfilling the past.

With our present organization, which must be continued, his connection with design has not been sufficient and he must, therefore, exert an extra effort to be thoroughly familar with all phases of it and understand the problems involved. He should not hesitate in presenting his ideas and should not wait until the testing is started to point out the errors. He should understand that it is his responsibility to develop a finished product on schedule and, therefore, his responsibility to see that the schedule is met, even when certain duties in connection with it have been assigned to other divisions. The Experimental Shop carries the responsibility for manufacturing and acquiring the material for his project. However, in case of a delay, it is his responsibility to obviate this if possible and to exert every effort to overcome the handicaps which will always be present. The point is that the Experimental Shop, similar to the test facilities, Installation Department, Purchasing Department, or similar group, is simply a required organization set up to facilitate his work and it always remains his responsibility to come out with the answer at the proper time.

The Project Engineer should have a thorough and up-to-date picture of the sales situation at all times. This is using the term "sales" in the broad sense. That is, the performance of the engine has to be satisfactory in relation to competition, or there can be no sales. Type tests, ATC tests and other official demonstrations are purely sales efforts, although considered engineering functions, and are conducted for the sole purpose of selling engines. Conformation to Army and Navy standards exists for the same reason. The Project Engineer should, in addition to being thoroughly familiar with the actual contracts, type tests and other scheduled demonstrations, visualize for himself and plan ahead on the required tests to make his project salable. On demonstration and semi-production engines he should be familiar in every detail with the orders, including delivery dates, performance, airplane demonstrations, and all conditions surrounding them, and it remains his final responsibility that the demonstration is a successful one.

On the manufacturing side, his first duty is obviously delivery of the bill of material to the Production Shop. It is his primary responsibility to see that his project gets through the Production Shop properly and is properly tested. It is his responsibility to plan ahead of time for any special equipment required by Production for testing and to see that the proper people are notified that this equipment is necessary. The organization which is set up under Campbell to take care of production troubles is an extra service which has been organized for the convenience of the Project Engineers and does not relieve him of the responsibility of seeing that his development is properly and expeditiously manufactured. In the same way an organization has been built up under Gove to design the test equipment for Production. This organization, however, does not relieve the Project Engineer of his responsibility for initiating at the proper time the procurement of the necessary equipment.

The matter of quality includes a great many things other than inspection proper. In the first place, the Project Engineer should be certain before the delivery date arrives that his development will meet its guarantees. This includes all specifications under which the item has been sold, but obviously the most vital of these are performance and weight. First, he, of course, has his own experimental check. There are then both the production and customer's checks on his guarantees for which he still remains responsible. In other words, as an auxiliary service, the Installation Department, or field service group, may follow certain tests in the field, but it remains the Project Engineer's responsibility to see that the guarantees are properly met.

His responsibility for service operation are concerned pretty much with seeing that his items are properly operated and then of obtaining remedies for troubles which may be encountered. The Installation Department obviously comes under the matter of correct operating conditions and because of its complexity it has been found necessary to have the Installation Department assume the responsibility for the details of the installation. However, again if this does not work properly it is the Project Engineer's responsibility to see that it is fixed and he should, therefore, be seriously concerned with the Installation Department. Because of the service they render him, it is necessary that he furnish them as far in advance as possible with all requisite information pertaining to the installation including the data required for such auxiliaries as oil coolers, radiators, etc.

I have written this summary in general terms in order to make it applicable to all Project Engineers. It is probably important, however, to have a complete understanding of responsibilities where the work overlaps. A good example is the case of supercharging. The Supercharger Project Engineer is a service to be called upon exactly as we would consult General Electric, or any outside concerns. If a supercharger is required for an engine design, it is the Supercharger Project Engineer's responsibility to furnish the necessary figures for this and wherever possible to substantiate these figures with past data, wooden model tests, or any other feasible method. In other words, the Supercharger Project Engineer will be held responsible for supplying the correct data. This does not relieve the Engine Project Engineer from the responsibility of developing this supercharger and correcting any errors or deficiencies as quickly as possible. As an extreme example, he might even disagree with a design from the beginning and request that outside consultants be called in to determine its correctness.

In brief, therefore, the Project Engineer should carry the basic responsibility beginning with design and carrying over to proper service operation and this is the picture the Project Engineer should see. As noted above, he must, of course, do this within the boundaries determined by general policy by the organization set-up and by the instructions of his immediate superior. In other words, many of the above responsibilities will be automatically lifted from him in varying amounts and to a varying degree. However, all of the rest must be carried by him and no requirements should be assumed to be properly taken care of unless he has checked and found them to be so. The Project Engineer is responsible in every case except where he has reported to his immediate superior and been relieved of the responsibility in the specific area.

The only note of warning concerning the assumption of this responsibility is that the basic requirement when operating as part of a large organization is that the Project Engineer should be specifically cautioned about keeping his immediate superiors completely informed of any changes or revisions, preferably while these are being contemplated. It is absolutely essential that those in charge have information of the action contemplated or taken as otherwise with the large number of projects, those in general charge will lose touch. It is better to err on the side of too much information rather than too little. Obviously, the whole thing is a matter of judgment which the Project Engineer must acquire. However, a proper realization of the necessity for keeping those in charge informed, together with the present regulation which specifies that no changes can be made in bills of materials without the approval of the responsible engineers (Willgoos and Parkins) should be sufficient safeguard.

Finally, the engineering organization as set up for supervision and administration of the Project Engineers should be made clear, as there has been some misunderstanding of this. The so-called administration of the Project Engineers is a matter of bookkeeping, maintenance of working conditions and discipline and the centralization of records. The Project Engineer is responsible solely to the engineer in charge for proper conduct of his work. The engineer in charge is his superior to whom he is primarily responsible and to whom he must look for all instructions concerning his project, promotion and pay advancement. In other words, the engineer in charge is the sole judge as to whether or not the Project Engineer is satisfactorily performing his duties.

                          L. S. Hobbs

Pratt & Whitney Aircraft "Space Transport Engines" Booklet Forecast of the Future

SPACE TRANSPORT ENGINES

Pratt & Whitney Aircraft DIVISION OF UNITED AIRCRAFT CORPORATION UA

Over the past four years, Pratt & Whitney Aircraft has conducted a comprehensive study to define the propulsion requirements for the next generation of launch vehicles. As a part of these studies, which have covered a wide range of sizes and missions, the engineering mockup shown here was constructed to assist in component arrangement and integration, and shows what a 250,000-pound thrust space transport engine might look like.

To provide an economic space transportation system, it appears that the next generation of launch vehicles will be developed as reusable systems. Engines for these vehicles will be significantly different from the throw-away types in current use, and must possess many of the characteristics of the engines in use in today's jet aircraft. Stability over a wide range of operating conditions, variable thrust to permit vehicle control and ground checkout, time between overhauls measured in hours, and operational dependability must be coupled with high performance to meet the propulsion needs of a space transport system.

The engine design represented by the mockup, which has been designated RL20P-3, employs oxygen/hydrogen propellants. It produces 250,000 pounds thrust at sea level, has a nozzle expansion ratio of 20.5, and an exit diameter of three feet. It operates at a main chamber pressure of 3000 psia at full thrust; thrust is variable from 10% to 100%. By using traditional transport engine design practice, the initial time between overhauls (T.B.O.) is expected to exceed 10 hours. This T.B.O. will allow 75 to 100 flights before engine removal for overhaul.

The thermodynamic operating cycle of the RL20P-3 is the same as that employed in an afterburning turbojet engine. In a turbojet, the working fluid is air with fuel added in the main burner and afterburner. In the RL20P-3 cycle, the working fluid is hydrogen with oxygen added in the preburner and main chamber. The hydrogen, after passing through the fuel pump, is ducted to the preburner where some of the oxygen is added to produce a working fluid of hydrogen and steam at approximately 1300°F. This passes through the turbines and provides the power to drive the pumps. In the main chamber, the balance of the oxygen is added to complete the combustion.

The tandem combustion cycle improves the combustion stability margin by having one of the propellants in the main chamber injected as a warm gas. In the preburner, both propellants are injected as liquids. However, the preburner mixture ratio is low, with inherently wider combustion stability limits, and the resulting low temperature level allows direct application of jet engine combustion damping devices.

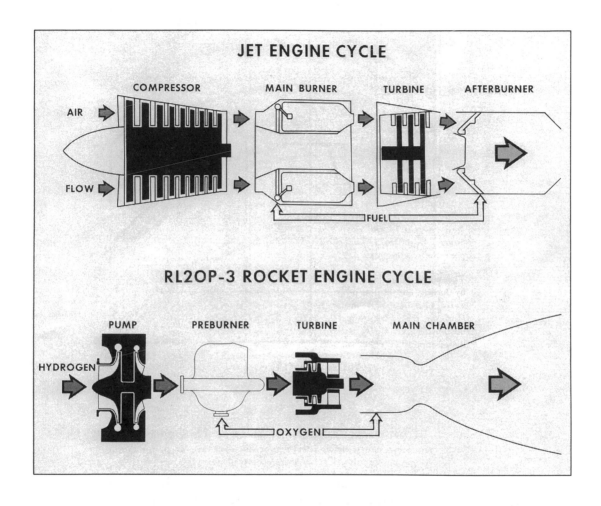

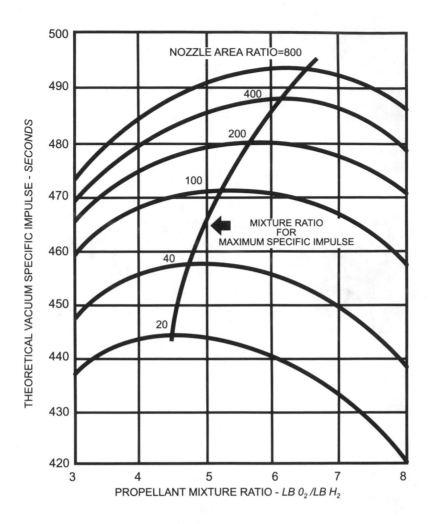

In general, for a particular thrust level and vehicle diameter, the effective nozzle area ratio is almost directly proportional to chamber pressure. Improved specific impulse (I_{sp}) can therefore be obtained by increasing the chamber pressure. A further advantage results from the higher chamber pressures because the mixture ratio for maximum I_{sp} increases with area ratio. The resulting higher propellant bulk density, in turn, reduces vehicle tankage weight.

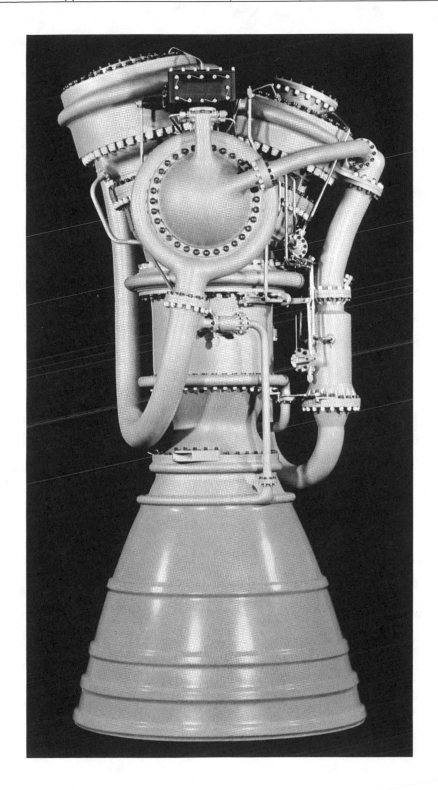

PWA FR-794 20 September 1963

Printed in U.S.A. 5M

Reviews Requested by NASA, Including Thank-You Letter From Bill Lucas

PRATT & WHITNEY AIRCRAFT GROUP

East Hartford, Connecticut 06108

7 November 1977

Dr. W. R. Lucas, Director
Code DA01
Marshall Space Flight Center
Alabama 35812

Dear Dr. Lucas:

As you requested, we have reviewed the design of the high
pressure oxidizer turbopump. Both the people at Marshall
and at Rocketdyne have been most cooperative in providing
us with a comprehensive overview of the design and the
problems encountered thus far. Since time constraints do
not permit us to make a detailed analysis of this specific
design, our comments are based on the experience gained in
the development of turbopumps with similar requirements at
our Company. The people on our review team were directly
involved in the design and development of three generations
of high pressure turbomachinery at Pratt & Whitney Aircraft.
Their report is enclosed for your information.

Sincerely,

UNITED TECHNOLOGIES CORPORATION
Pratt & Whitney Aircraft Group

R. C. Mulready
Director, Technical Planning

RCM:ams
Attachment

UNITED
TECHNOLOGIES

REVIEW OF SSME
OXIDIZER TURBOPUMP

PRATT & WHITNEY AIRCRAFT GROUP
REVIEW TEAM

J. T. Akin
W. E. Creslein
W. A. Ledwith
H. J. McLean
R. C. Mulready
J. J. Sandy
W. E. Poole

November 3, 1977

TURBOPUMP CONFIGURATION

The basic turbopump concept selected for the SSME design appears to have several inherent weaknesses including:

1. Double suction impeller

2. Lack of separate thrust balance piston

3. Both bearing sets on LOX side

4. Internal bearing coolant supply and

5. Inability to balance as an assembly at high RPM

The selection of the double suction impeller, while attractive from a weight and efficiency point of view, results in certain inherent risks for this turbo-pump. It leads to high shaft RPM, small shaft and bearing diameters, relatively long pump length and the potential for high impeller side loading. These factors have produced a design which has three critical speeds in the operating range and in combination with (3), results in a rotor system which is very sensitive to un-balance. Balancing at low speed is not effective because couples that increase unbalance appear when the shaft shape changes at the higher modes. In fact, balance weight adjustments made at low speeds may increase the unbalance at higher speed.

The thrust balance system such as that used in the current design is sensitive to both axial and radial gaps at the critical throttling points and the shaft de-flections which occur with rotor bending tend to introduce dynamic hydraulic loads in both the axial and radial direction. These forces may contribute to the "400 Hertz" problem.

By using a separate piston for thrust balance, as shown in Figure 1, it is possible to design a very stiff system, less subject to rotor and housing de-flactions and to decouple the effects of changes in radial gap which inevitably occur in the impeller and housing systems. This can probably not be accomplished with the long rotor resulting from use of a double suction impeller. The double suction design shown in Figure 2 offered a 135 pound weight saving over the single suction design in Figure 1. Excessive rotor length prevented the use of a separate thrust piston in this design.

The decision to put both bearing sets in the LOX side has forced approximately 50% of the rotor mass to be overhung. This leads to an extreme sensitivity to turbine balance. The system will not withstand the potential loss of a single turbine blade.

The bearing coolant flow to preburner pump bearings is supplied by the leakage through a very high pressure drop labyrinth seal with close clearances which makes it subject to large variations in flow and pressure drop across the bearings in the event of a rub. The bearing coolant to the turbine end bearings, supplied through the center of the shaft, has potential alternative leakage paths around the bearings. A separately metered externally supplied coolant flow would be more reliable and would allow bearing coolant to be supplied for high speed balancing of the complete assembly if the problem of shaft shape changes at high speed can be overcome.

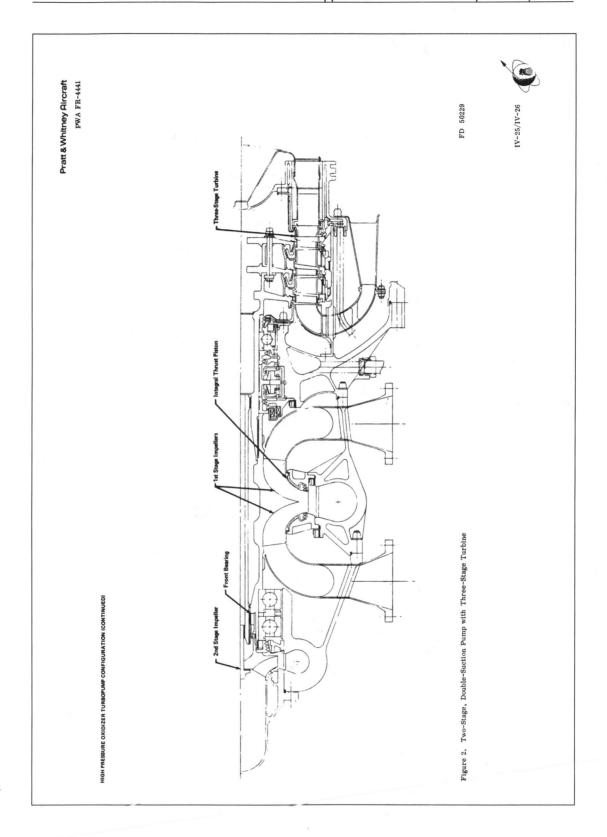

Pratt & Whitney Aircraft
PWA FR-4441

HIGH PRESSURE OXIDIZER TURBOPUMP CONFIGURATION (CONTINUED)

2nd Stage Impeller

Front Bearing

1st Stage Impellers

Integral Thrust Piston

Three-Stage Turbine

FD 50229

IV-25/IV-26

Figure 2. Two-Stage, Double-Suction Pump with Three-Stage Turbine

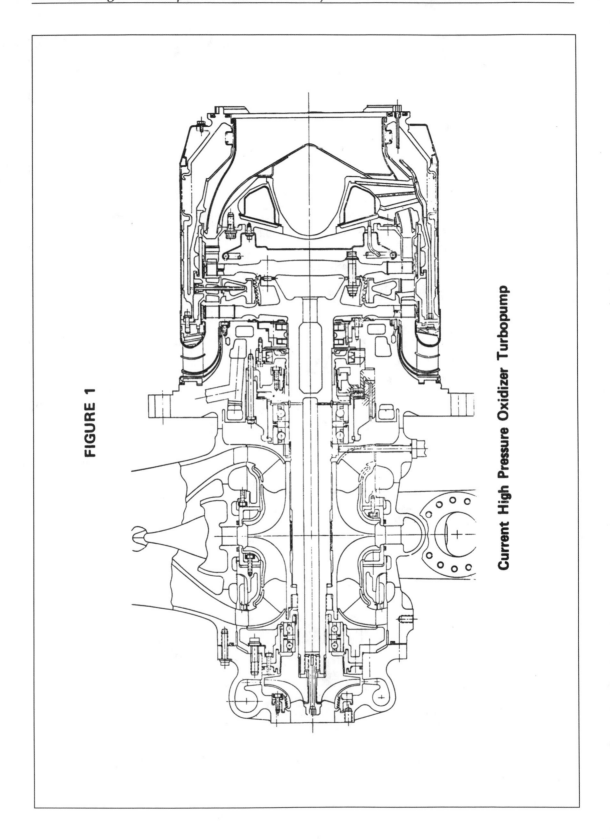

FIGURE 1

Current High Pressure Oxidizer Turbopump

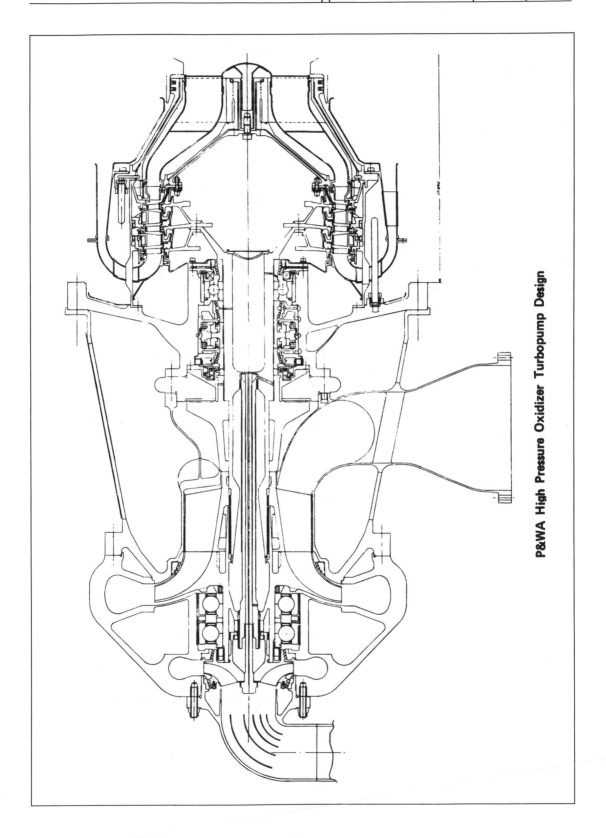

P&WA High Pressure Oxidizer Turbopump Design

ROTOR DYNAMICS

SUMMARY:

The bearing distress encountered in the turbopump is probably due in large part
to the magnitude of the radial loads which are being produced by synchronous
critical speed response. The non-sychronous response encountered at high speeds,
although not explained in detail at the briefing, does not appear to be a major
contributor. The "400 Hz" mode identified as a case resonance is in reality a
coupled rotor/case mode, highly sensitive to rotor imbalance in the turbine area.
Present balance limits are too lenient and the low speed, rigid body technique
is marginal due to the flexure of the rotor at design speeds. Turbopump reliability may require major changes relative to:

- o Shaft stiffness
- o Bearing radial load capacity and stiffness
- o Turbine overhang
- o Balancing procedure

DISCUSSION:

The nature of the pump critical speed characteristics and balancing techniques
are such that extremely high radial bearing loads can easily be obtained during
pump operation. According to Dr. Gunter's analysis, three critical speeds have
to be traversed before reaching FPL as illustrated in Figures 3 and 4. The first
mode is pure rigid body motion of the rotor and case and is of no concern. The
second and third modes, however, involve bending of the rotor and translation of
the turbine which make the response at these speeds sensitive to imbalance,
especially in the turbine overhang. The fourth mode is of no concern since it
is out of the operating range.

The sensitivity of radial bearing loads to imbalance is illustrated in Figure 5.
Loads of approximately 3,000 lbs. are seen at the 2nd and 3rd critical speeds
for 8 gm-in turbine imbalance. With present balancing techniques, this magnitude of imbalance in the rotor is not unrealistic due to:

- o Assembly of pump after final balance
- o Thermal and centrifugal shifting of parts during
  turbopump operation
- o Rotor/case rub during turbopump operation
- o Torsional windup of shaft tending to create
  imbalance
- o Possible formation of ice in shaft cavity
  adjacent to turbine
- o Possible non-uniformity in bearing coolant flow
  in shaft
- o Uncertainty in predicting fluid imbalance in
  impeller

The imbalance sensitivity is further aggravated by the bending in the rotor at the 2nd and 3rd critical speeds which is not present during the low speed balancing operation. This bending is controlled primarily by shaft diameter, bearing span, turbine overhang and bearing/housing stiffness.

It does not appear that the turbopump has a classic instability problem, although the fact that non-synchronous vibration occurs at the higher speeds indicates that destabilizing forces are present and need to be eliminated. The cause of the non-synchronous vibration does not appear to be understood based on data presented at the briefing. The fact that pressure fluctucations at the vibration frequency are seen in the impeller volute suggests a hydro/ mechanical coupling that could produce instability.

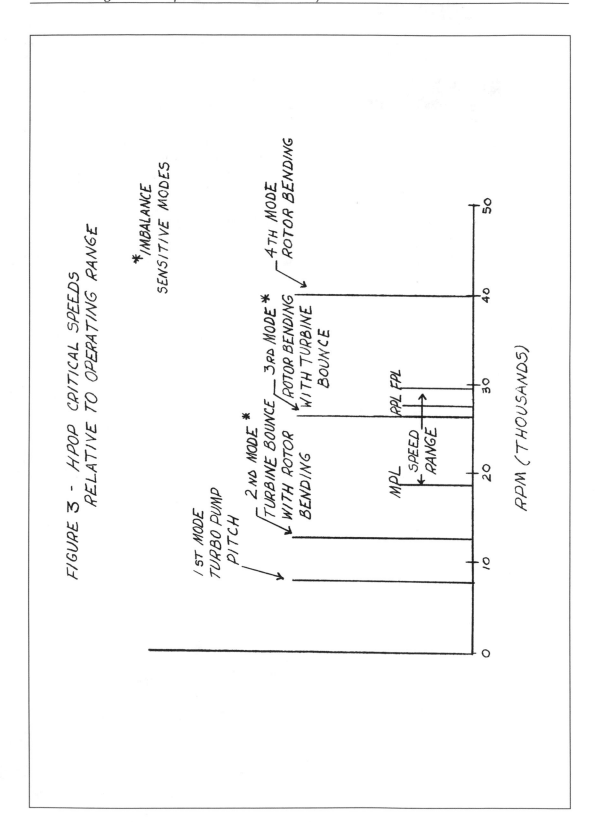

FIGURE 3 - HPOP CRITICAL SPEEDS RELATIVE TO OPERATING RANGE

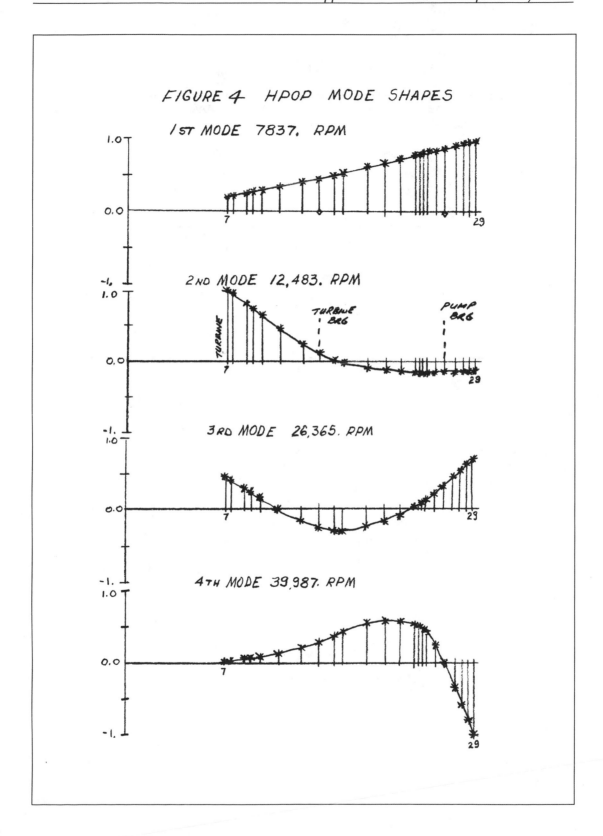

FIGURE 4   HPOP MODE SHAPES

FIGURE 5

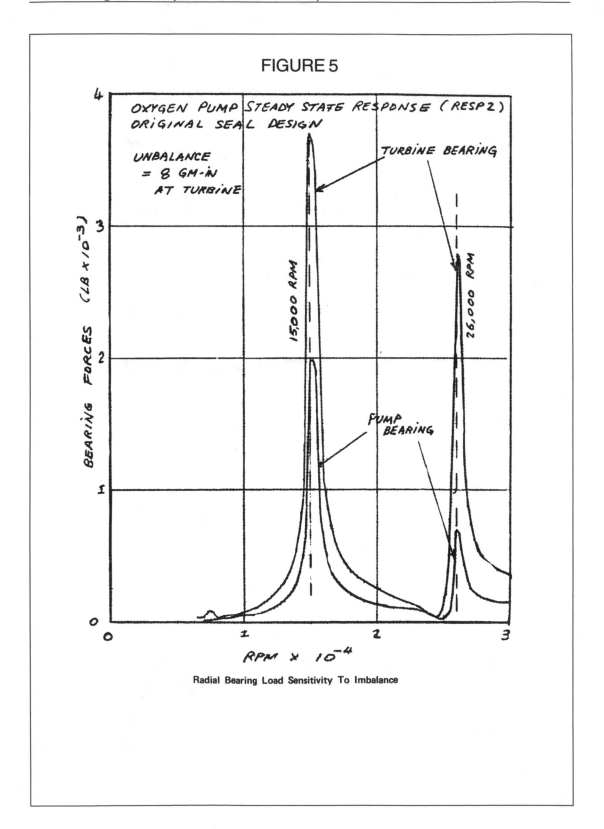

Radial Bearing Load Sensitivity To Imbalance

LCF CALCULATIONS AND LOCAL STRESS

There are several areas in the HPOTP that have high local stresses, particularly at pump exit vanes and the transition section between pump and turbine housing. The work that was done to identify these high stresses was performed with finite element analysis. The loads applied to the cases were from pressure and installation loads at steady state temperature.

LCF calculations based on the coarse finite element breakup shown in the briefing are not indicative of the true surface local stress that can lead to early rupture. We have found as much as a 2 to 1 difference in predicted surface stress with different finite element size. In light of the low LCF lives quoted on major structural parts with the element breakup shown and the absence of transient thermal stress analysis, we are concerned that the life quoted may be optimistic.

It is our design practice to avoid splines and threads in disk bores. Disk growth and angular deflections produced by torque loads tend to cause high local stresses in the roots of splines and threads. LCF and crack propagation considerations can be critical in a design with stress risers in high deflection areas such as disc bores. We believe that this is a particularly significant risk in a cast material with low elongation.

HOUSING DEFLECTION

The HPOTP review did not address housing deflections. Housing deflections as well as rotor deflections must be known to provide a workable design. The major concern is the effect of radial and axial deflections at seals, bearings and thrust piston orifices.

Both the thrust piston impeller and mating surface on the housing will have radial and axial deflections at various power level settings. The housing restriction surface will have a distorted circumferential pattern due to the vane axial restraints around the volute and the varying size of the volute collecting system. It seems as though unsymmetrical flow into the thrust piston will occur causing unknown pressure loading, both radial and axial.

MAIN IMPELLER STACK

The short coupled main impeller stack is very sensitive to preload control. The inherent high spring rate is intolerant to a slight angle-of-turn of the nut, which can occur with the tablock devices. We consider such an axial loading approach to be unreliable on the device intended for axial load balance. Should the nut loosen, even within the confines of its lock, the impeller may be given enough freedom to move back and forth on the shaft. This creates the potential for fretting, heat generation, and fire hazard, as well as a dead band in the thrust balance system. Although it was reported that no signs of such fretting has been seen, it has been our experience that such configurations (short stack) should be avoided. It would appear that a long stack tie bolt might be used which would also allow significant stiffening of the rotor by using the larger diameter of impeller, bearing and seal elements.

THERMODYNAMICS

The temperature breakdown shown for the turbine end of the turbopump will produce ice inside the shaft. While not a problem initially, with repeated use enough water could condense in this cavity to create shaft unbalance. The long turbine overhang is not tolerant to any shaft unbalance.

The analysis presented in the briefing assumed adiabatic wall conditions inside the shaft which appears to be incorrect since condensation and freezing are likely to occur in that area. No transient thermal analysis appears to have been done. The affects of thermal strains on the turbopump structure could be significant, particularly in view of the low LCF lives quoted for the major structure.

FLUID MECHANICS

It has been our experience that high pressure turbomachinery has little margin for errors in the prediction of axial and radial pressure balance.

Several features in the design where pressure mal-distribution can occur were brought out in the briefing with apparently no verification by test measurements of the hardware. These features are:

o Side entry on the main impeller, even when carefully executed, can lead to flow inlet distortion and consequent discharge pressure distortion.

o Single discharge volute design, even with vaned diffuser, can create a load bias.

o Pump impeller flowpath non-concentricity can lead to discharge pressure variations around the impeller circumference - a synchronous load unbalance cause by the "best pumping" passage.

o The thrust piston side loads were not accounted for. Shaft and housing deflections will cause unequal orifice gaps around the circumference of the impeller with attendant pressure fluctuations - possibly a source of hydro-dynamic instability.

With the potential sources for hydraulic loading well beyond those assumed in the bearing load analysis, it would appear that test verification is most important. We recommend that a pump be fitted with close-coupled pressure transducers around the periphery of the volute and around the thrust piston cavities to assess the magnitude of the hydraulic load unbalance.

H.P.O.T.P. ASSEMBLY PROCEDURE

The assembly procedure of the HPOTP creates unknown risk with regard to successful turbopump operation. The blind assembly of the preburner impeller subassembly, that contains the preburner impeller, preburner ball bearings and preburner pump impeller housing, could be the source of significant undetectable damage. Many things could occur in this blind assembly and the inspection and checks are limited. Lack of visual inspection of the ball bearings and lack of on-assembly balance of turbopump create risk. The turbopump goes to test with little confidence of proper impeller seating and loading.

Bearings are assembled on the impeller with a tight fit and then the shaft is assembled tight to the impeller creating more radial deflections in the brittle 440C bearing inner races. If a crack or damaged inner race results due to the blind assembly, it is not known until the turbopump test.

Other areas of concern are the chance that symmetrical parts could be assembled backward. The angular contact bearing races or the turbine bearing jet could be reversed at assembly.

BEARING LIFE REVIEW

BEARING LIFE

The bearing sections selected appear to be much too light for the service. It is unlikely that Design life of 7.5 hours before pump overhaul can be achieved with present bearing system. High loads and low capacity bearings will limit theoretical fatigue life and premature failure from other failure modes are expected.

Predicted radial loads are excessive and result in unacceptable thrust/radial load ratio. This is expected to result in circumferential pocket wear and O.D. (guide surface) wear both of which are evident now in pump run bearings. In addition, wear into the fiberglas reinforced cage is expected to result in ball abrasion and wear. In our 350K pump work, we found Salox-M to have better wear characteristics than Armalon for ball bearings. The observed anti-rotation lug wear and fracture is indicative of high rotating radial load.

Bearing internal geometry appears to be marginal for a non-lubricated, cryogenically cooled application. The close inner race curvature will result in high local heat generation at the ball/race contact. This can occur locally within the contact even when total cooling is ample. This high local heating will degrade race and ball surfaces leading to premature failure. Severity of this problem is aggravated by the coarse cross race finish. Ball and race banding is common even on low time pumps.

Shaft/housing misalignment effects appear not to have been considered. Mis alignment in ball bearings results in increased ball excursion. This can produce cage wear similar to high radial loads and in extreme cases can load balls locally over race shoulder and down to the race bottom. Another effect of misalignment in duplex bearing applications is the internal moment load which further decreases life.

BEARING COOLANT

Bearing coolant management is not well controlled. Use of seal leakage as bearing coolant creates high hydraulic load which unbalances preload sharing between bearings compounding the total load and thrust/radial load ratio problem. Seal deterioration, from any cause, will accentuate the load sharing problem and can precipitate early bearing failure.

The selected flow path is opposite the direction of bearing pumping resulting in unnecessary churning heat generation and excessive flow resistance. The high resulting pressure difference causes cage axial loading with consequent side pocket wear which was observed and the potential for ball abrasion.

The downstream bearings (#2 or 3) are penalized with upstream bearing heat and the possibility of 2-phase coolant which may result from the high local heat at the race contact.

The proposed turbine end coolant revision would not appear to solve the problem. The intent is to provide an axial jet but centrifugal effects resulting from jet rotation may override. The lack of "foolproof" assembly may pose future problems.

BEARING LOADS

Actual magnitude of bearing loads is in doubt due to uncertainty of hydraulic side loads and turbine balance sensitivity. The high sensitivity to turbine unbalance results in no life margin for balance deterioration.

Load estimation by back calculation from bearing running path measurements is inaccurate. Tolerance stack-up and multiple paths from different operating conditions combine to create potentially large errors.

BEARING TEST SUBSTANTIATION

Rig testing as performed to date is of little value because rig conditions don't simulate pump operation, especially the detrimental effects of high radial load. Radial loads have not been applied in any rig testing to date.

PRATT & WHITNEY AIRCRAFT GROUP

East Hartford, Connecticut 06108

26 January 1978

Dr. W. R. Lucas, Director
Code DA01
Marshall Space Flight Center
Alabama 35812

Dear Dr. Lucas:

On January 11th and 12th, we reviewed the high pressure
fuel turbopump turbine blade failures and associated turbine problems. Your team from Marshall and the group from
Rocketdyne provided comprehensive presentations covering
all facets of the problems. As with our review of the
oxidizer turbopump, other commitments have precluded a detailed analysis of the information presented. Instead,
our observations, contained in the attached report, are
based on the experience gained in the development of similar equipment at our Company.

If you or your people have any further questions please
feel free to contact me.

Sincerely,

UNITED TECHNOLOGIES CORPORATION
Pratt & Whitney Aircraft Group

R. C. Mulready
Director, Technical Planning

RCM:je
Attachment

cc: Mr. J. Thomson
  Code EE21
  Marshall Space Flight Center
  Alabama 35812

Mr. J. Hager AC 38
Rocketdyne
6633 Canoga Avenue
Canoga Park Calif. 91304

UNITED
TECHNOLOGIES

REVIEW OF SSME

FUEL TURBOPUMP TURBINE

PRATT & WHITNEY AIRCRAFT GROUP

REVIEW TEAM

J. T. Akin
R. R. Atherton
A. Hauser
W. A. Ledwith
H. J. McLean
J. W. Morris
R. C. Mulready
R. R. Sellers
C. B. Stevens

FIRST STAGE TURBINE BLADE FAILURE
MAIN FUEL TURBOPUMP

From the presentations made by NASA-Marshall and Rocketdyne on 1/11 and 1/12, we conclude that failure of the first stage turbine blade on test number 902-095 was caused by high cycle fatigue in the airfoil root. The information presented identified the typical striations associated with high cycle fatigue in the fracture face. Our metallurgists identify the fracture as typical of high cycle fatigue in directionally solidified nickel base super-alloys.

In addition to the turbine blade problem, several other areas of distress in the turbopump were discussed, including bradalloy tip seals, housing thermal fatigue, housing heat shield cracking and nozzle vane erosion. These problems appear to be understood, and corrective actions are being explored.

Our review of the design approach and the turbine environment revealed a number of factors which would increase the likelihood of failures in the high cycle fatigue mode. These factors include:

1. The potential for a strong 13E excitation source from the turbine inlet bearing support struts which does not appear to have been addressed in the design analysis. The current design is sensitive to this excitation in both first torsional and first bending modes of vibration.

2. There is evidence of blade damper lock-up and possible platform interference between blades, making the dampers ineffective. The high relative stiffness of the blade necks may also preclude effective damper action.

3. The blade has a brittle coating of ZrO_2. Since this coating is less ductile than the underlying blade material, there is a high probability of surface crack initiation by low cycle thermal fatigue or high cycle vibratory fatigue.

4. The lack of an experimentally verified Goodman diagram and measured steady and vibratory stresses make an assessment of the design's adequacy difficult. The calculated steady stresses are so high as to allow very small vibratory stress levels (about 5,500 psi, zero to peak).

5. The turbine inlet temperature is running some 450° over design.

Our general impression is that although an extensive analysis has been made of the design, some significant elements are missing including local stress analysis of the airfoil root platform area and recognition of the potential 13E excitation. In addition, there has been no experimental verification of stress levels, nor is there adequate material data to statistically ascertain the capability of the blade to live in its assumed environment. The work which we deem necessary to define the problem is contained in the following paragraphs.

1. <u>Spin Pit Test</u>

   A spin test is necessary in order to accurately determine the local steady
   stresses in the airfoil root section without the influences of gas bending or
   thermals. Strain gages located at the airfoil leading edge, trailing edge
   and maximum thickness and any other critical location (as identified by
   finite element analysis) should be used. This spin test would be used to
   calibrate the finite element analysis results.

   Simulated tests such as pulling on the blade with mechanical means are not
   considered sufficiently accurate to accomplish this purpose.

2. <u>Strut/Vane Excitation Potential</u>

   Flow-induced wakes from struts and vanes both upstream and downstream of the
   blades are potential forcing functions for blade vibration. The thirteen
   turbine inlet bearing support struts are probably the largest driver and
   apparently were not considered in the design analysis.

   A Fourier analysis of the strut/vane wakes will identify potential high
   vibratory stress levels within the operating range of the pump when plotted
   on a Campbell diagram as illustrated in Figure I. Pratt & Whitney's approach
   is to identify all known excitation sources, particularly the low orders, and
   design to avoid them. Rocketdyne's current design appears sensitive to 13E
   in both first torsional and first bending modes.

3. <u>Engine Dynamic Strain Gage Test</u>

   An instrumented engine test is necessary to determine the true vibratory
   loading on the blade and accurately define any corrective action. This test
   can be used to optimize the size of the dampers by measuring dynamic stresses
   on blades with different size dampers. Dynamic gages should be placed at the
   airfoil leading edge (failure location) and other critical locations as determined by the blade mode shapes.

   Overall and discrete frequency alternating stresses measured as a function
   of RPM should be plotted on an experimentally determined Goodman diagram to
   determine the blade capability. Steady stresses on the Goodman diagram must
   be determined from a finite element analysis properly calibrated from spin
   pit results, since the critical stress will probably be localized.

4. <u>Local Stress Analysis</u>

   It appears that a "fine mesh" finite element local stress analysis was not
   performed on the blade. When blade failure occurs, it will originate at a
   point of high local stress, whether caused by surface anomaly, thermal gradient,
   or geometric configuration. It has been our experience that a three-dimensional
   finite element analysis of the NASTRAN variety is required to identify areas
   of high local stress. Particular attention should be paid to airfoil/platform
   junction, platform fillet radii into the shank, load path of the leading and
   trailing edge, and the fillet from the shank into the attachment. The analysis
   and design should be iteratively manipulated until local stresses are reduced

to an acceptable level. The results of the analysis can also be used as
a guide to the acceptability of parts produced with porosity, dross, etc.
The analytical results should be experimentally verified. (See recommendation
#1).

5. Goodman Diagram

Allowable combinations of steady and vibratory stress must be determined
relative to an accurate Goodman Diagram. A straight line assumption based
on endurance limit and ultimate (or yield) strength is not adequate for
nickel-base superalloys whose Goodman Diagrams have been found to deviate
greatly from straight lines (Figure 2). The Goodman Diagram must be generated
experimentally by applying a combination of steady and vibratory stress to
the test piece.

Test temperature must be the turbine blade local operating temperature and
a coating must be applied if it is to be used in service. It is recommended
that actual blades be used as the test pieces to assure exact duplication of
material characteristics. The number of test points generated at each stress
condition should be sufficient to allow a statistical projection to a failure
rate acceptable for the design. Statistical treatment is necessary in a
turbine containing several hundred airfoils produced by investment casting
which produces a considerable scatter in material properties.

6. Blade Coating

It is our opinion that the coating should be removed from the turbine blades
since it appears that the detrimental effect of airfoil coatings may out-weigh
any advantage for this application. Coatings developed for turbine blades are
intended to provide oxidation/corrosion resistance (NiCrAlY) or thermal insulation (ZrO_2).

However, these coatings are far less ductile and weaker than the underlying
Nickel-base superalloys used for turbine blades (particularly directionally
solidified alloys). As a result, the addition of a coating tends to increase
the probability of surface crack initiation by both low cycle thermal fatigue
and high cycle vibratory fatigue. Experience has shown that a surface crack
may initiate due to the large local strain experienced during an engine thermal
cycle and then propagate through the blade as a result of high cycle vibratory
stress.

Recent heat transfer experiments conducted by PWA and NASA (Frank Stepka) have
shown that the insulating characteristics of a 0.010" layer of ceramic coating
such as ZrO_2 may be offset to a degree by the higher convective heat transfer
coefficients which result from the characteristic porous surface roughness of
these coatings. (Even in a polished state surface porosity can produce an
effective roughness.) Using a 0.005" thickness may in fact result in a hotter
metal temperature than the uncoated blade would other wise experience.

In addition, blade coating has little or no load carrying capability and,
therefore, becomes a "dead load" that increases critical blade root centrifugal
stress.

7. Blade Redesign

We expect that the increased steady and vibratory stress, coupled with an experimentally verified minimum property Goodman Diagram and local stress analysis, will show the blade design to be unacceptable. It is recommended that a blade redesign be started. In the absence of engine dynamic data, steady stresses in the redesign should be low enough to allow a 15,000 psi zero to peak vibratory stress. As a first cut, we would limit tensile plus restored gas bending to 75% of 0.2% yield strength. The existing design looks good from a load path standpoint, although the leading edge and trailing edge cracking problems may be aggravated by the stiffness of the shank under the leading and trailing edges. The apparent relatively high stiffness of the shank may also limit platform motion and reduce the effectiveness of the damping system. To meet the above criteria, the blade may have to be cored to reduce the pull stress to an acceptable level. This would have the added benefit of reducing thermal strains which lead to airfoil cracking.

8. Damper Redesign

The dampers do not appear to be functioning properly. They appear to be causing blade lockup. As a first cut, the damper can be sized by calculating the damper load which will allow slip with a 0.5 coefficient of friction when the bending stress in the blade shank is at half the allowable. How ever the damper is sized, it should be experimentally verified (see recommendation #3).

Regarding the damper design, the contact points between damper and platform seem highly indeterminate due to tolerances and surface coatings on both parts. The damper should be designed preferably for a three-point contact (two on one blade, one on the adjacent) to assure operation to design intent. As damping depends to some extent on damper spring rate, the contact points should be placed to maximize the spring rate.

A recheck of the platform gaps seems to be in order. Curved platform sides make platform gaps sensitive to axial positioning. Although the blades may all be located against their forward stop at assembly, subsequent handling may cause some blades to move forward against the rear cotter pin retainer, closing the platform gap and rendering damping ineffective. Since the damper serves as a platform gap seal, opening the gap should have little deleterious affect.

The design should primarily address damping the first bending mode, and secondarily address first torsional mode, assuming there is enough platform participation in the torsion mode to make the damper effective. It is doubtful if the damper, or any damper, would be of any value in attenuating stresses at the higher order modes, such as the 41E associated with the first stage vanes.

9. Temperature Profile

It would be most desirable to experimentally determine blade metal temperatures. We would investigate the applicability of the advanced "two-color" optical pyrometry for this turbine. If metal temperatures can not be measured directly, then it is necessary to measure gas temperature entering the turbine. The blades average the circumferential temperature gradients, resulting in a single characteristic radial temperature profile experienced by all blades. To measure this profile experimentally, the gas temperature should be sampled at a number of radial and circumferential locations. It is recommended that a minimum of 50 thermocouples be extended from the 1st vane or bearing support strut leading edges (5 radial locations) to provide these temperature readings during engine operation.

10. Turbine Inlet Temperature

All of the other observations will be of little value unless the turbine inlet temperature is reduced to more nearly original design values. Pratt & Whitney was not informed of the detailed plans to improve turbopump performance and hence can not make a judgement on the probability of success.

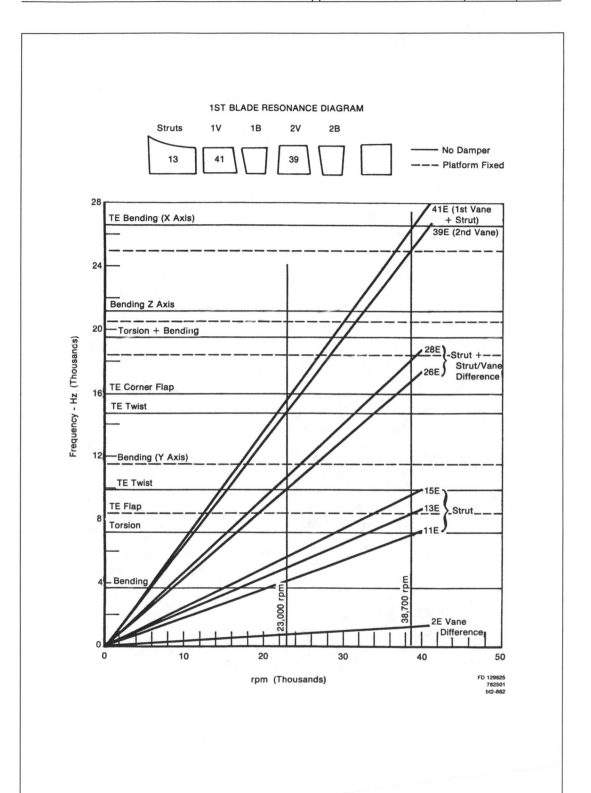

1ST BLADE RESONANCE DIAGRAM

FD 129625
782501
bt2-662

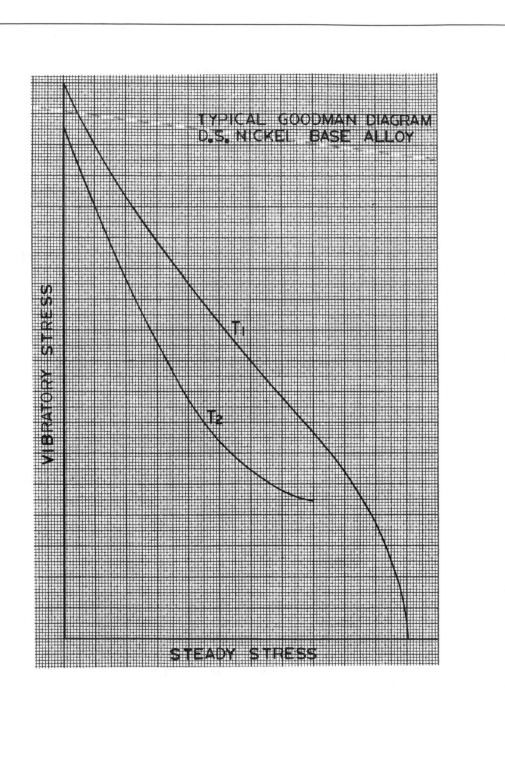

National Aeronautics and
Space Administration

George C. Marshall Space Flight Center
Marshall Space Flight Center, Alabama
35812

R. C. MULREADY

DEC 5 1977

NOV 2 8 1977

Reply to Attn of: DA01

Mr. Richard C. Mulready
Director, Technical Planning
United Technologies Corporation
Pratt and Whitney Aircraft Group
East Hartford, CT 06108

Dear Dick:

I want to express my personal thanks to you and your review
team for an excellent job in the review of the SSME high
pressure turbopump design. Your participation in meetings
at Marshall and Rocketdyne has been most helpful, and your
excellent report to me and my staff on November 8 will be
fully utilized. I believe a significant contribution has
been made already to the program by your efforts, and I'm
confident that other benefits will accrue as we are able
to implement additional suggestions. The quality of your
efforts in this instance is entirely consistent with the
high appraisal of Pratt and Whitney that we at the Marshall
Space Flight Center have always held.

The highly professional manner with which you conducted
the review and the technical expertise displayed are a
credit to you, your corporation, and the aerospace industry.
Those of us who have been privileged to work with the nation's
aerospace industry know and appreciate the spirit of coop-
eration and teamwork which has characterized the relationship
between NASA and the industry since the beginning of the space
program. Your performance in the SSME turbopump design review
is proof of the continuation of that fine spirit.

Please convey my appreciation to all of your team.

Sincerely,

Bill Lucas

W. R. Lucas
Director

cc:
Mr. Harry Gray
Mr. Bruce Torrell

Index

Abbreviations are used after the page number to indicate a figure (*f*).

High-pressure rockets *(continued)*
    Pratt & Whitney design review of SSME, 147–148
    reusable engine brochure, 93
    RL20 mockup, 92*f*, 92–93, 94*f*
    staged combustion engine
        development work, pre-1969, 131–132, 132*f*
        lack of Pratt & Whitney commitment, 145–146
        NASA's preference for, 134–135
        rocket test area, 133, 133*f*
Hobbs, Luke, 9, 10*f*, 160*f*, 161–163
Horner, Jack, 54
Horwath, Al, 144
Howmet Corporation, 154, 155
Humphrey, Hubert, 99
Hydrogen. *See* Liquid hydrogen fuel

INCO 718, 112
Intelsat, 61
International Air Transport Association, 135
Irvine, General, 54
ISH, 113

J2 Apollo engine, 93, 94*f*
J42, 9
J57 (JT3)
    development of, 9
    dual-spool engine basis, 20
    modifications for 304 engine, 34–35, 35*f*
    size compared to T57, 19*f*
J58 turboramjet, 56
J75, 26
Jennings, J.C., 78
Jet burner test stand, 8
Johnson, Herrick L., 32–33
Johnson, Kelly, 29, 55
Johnson, Lyndon B., 61
Johnson, Roy, 57
Johnson, Stan, 144
Journey to Tranquility, 145
JT3D, 26

About the Author

(Courtesy of John Robson)

Richard C. "Dick" Mulready spent almost four decades developing advanced engines at the Research Laboratory and the Pratt & Whitney Aircraft division of United Aircraft during the exciting era when development of the staged combustion high-pressure engine transformed the Space Shuttle from idea to reality. He was the first project engineer for the RL10 liquid hydrogen rocket engine which, over the last 45 years, has launched most large satellites and space vehicles.

Dick was born on July 15, 1925, and graduated from Massachusetts Institute of Technology in February 1946 as an aeronautical engineer (engines option). He attended college under the auspices of the U.S. Navy V12 program and was mustered out in July 1946.

With a lifelong interest in aircraft engines and a favorable bent toward Pratt & Whitney, Dick's only college interview was with United Aircraft Corporation, of which Pratt & Whitney was a division. Dick joined the United Aircraft Research Laboratory in November 1946 and enjoyed a 37-year career there. The first 25 years of his career and the challenging state-of-the-art aircraft engine development projects on which he worked are the subject of this book.

Dick has authored or co-authored several technical papers, and he holds eight patents. In 1974, he shared the Goddard Award for his early work with liquid hydrogen rocket engines. Over the years, he has been a member of the Society of Automotive Engineers and the American Rocket Society, and an associate fellow of the American Institute of Aeronautics and Astronautics. His paper, "Space Transport Engines," found in Appendix B of this book, described at that time his view of what he believed the next generation of rocket engines would be. The paper was adopted by NASA for the Space Shuttle.

In addition to his love of aircraft engines, Dick is an avid fan of sailing. He admits, "For some people, cruising on a small boat is an enjoyable occasional pastime. For me, it is an addiction." Dick and his wife Carol have owned a variety of boats during their 45-year marriage and have cruised with their children along the coast of New England and through the Bahamas. In his retirement, Dick and Carol have sailed 18,000 miles from Maine to Florida to the Bahamas in his largest vessel, *The Leading Edge*, a magnificent craft with 34 feet on the waterline and a 12-foot 8-inch beam. Most recently, Dick completed his four-year restoration of a 70-year-old catboat, the *Sweetie*, which had been abandoned as a "basket of sticks" in his friend's boatyard. The 14-foot *Sweetie* now carries Dick and his grandchildren on sailing and fishing excursions.